TABLE OF CONTENTS

ACKNOWLEDGEMENTS

Thank you, God.

Thank you to my publisher, editor, creator, and all those I didn't meet but helped me help others. What a great crew!!!

Thanks, Mary Ann.

Thanks, Tony, Jordan, Corey, Walt, Mike and Jerry. You all believed in me when the most important people in a person's life didn't.

Thank you to those who were brave enough to tell others that they saw a ghost or paranormal entity and no one believed them. Time is on your side.

Many, many people on an investigation team spend countless hours watching videos, listening to audio recordings and checking information obtained from paranormal gadgets. After spending so much time and then not coming up with anything tangible it seems like a waste of time. It's not. There is no glamor or fame in the behind the scenes research. It can get boring. I appreciate you! Like most jobs in life, the less glamorous the job, the more worthwhile it is to humanity.

A MOMENT OF SILENCE

For the person who is being terrorized and harmed by paranormal entities, and doesn't have the information or resources, or who got told "no" when they asked for help,

keep asking.

I know it's hard to imagine, but everything happens for a reason.

You're an inspiration to the living while you deal with every fiber of your being with the ruthless dead.

CAVEATS

I'm trying to be as factual as possible. You can believe in whatever suits you. I'm not endorsing a particular cult, religion, political belief, ideology, or technology. The goal of any reasonable human being is not to deliberately harm others. I want all of us in the profession to communicate so we can make humanity better. I most positively don't know it all. If you have feedback on this book (positive or negative), I welcome it at: kirk_mcmanus@yahoo.com

PART 1

PROFESSIONAL STANDARDS OF THE INVESTIGATOR

Return this book to the publisher to if you want to read about cemeteries, skulls, knives, blood, gore, violence, people running scared, near misses, creepy places and spaces, worn-out dolls, things that go bump, handprints, and odd-looking faces. There are plenty of documentaries, online videos, social media sites, and books that will provide that. This book is from another angle.

YOU'RE NOT CRAZY.... FOR NOW

The first thing I tell my paranormal clients is that most likely, they aren't going crazy or losing their marbles. Paranormal "stuff" happens all the time. But one client, I had got several investigators in her home at different times, and the investigator saw the client putting injections into her eyes while the investigation team was there to make her eyes look bruised. Then the homeowner claimed she was being physically abused by an entity, and her bruised eyes proved it.

Then I tell them that our conversation is strictly confidential. Then they can open up without being judged. Then the flood gates open.

Our clients contact a paranormal investigator because they want someone outside of their social or work circles to validate that they aren't crazy. They don't want to spend several thousand dollars on paranormal investigation equipment. They don't want to spend copious amounts of time on the Internet or buying books to figure out a ghost. A coincidence started becoming a pattern. A rare event, the norm. Over time things got predictable, and along with that predictability came confusion, fear, worry, and doubt. The inner monologue of someone who is encountering a paranormal entity is one of overwhelming entrapment. I know. I've lived it.

I would like to be prolific and say that humans study the paranormal to "better humanity" and to understand the universe we live in.

But based on what I've heard, seen, and lived, most people don't study the paranormal as a first choice of a profession. They had a paranormal encounter and spent their time trying to figure it out, and eventually, studying the paranormal became a hobby that led to a profession, then a social media following, then a group of like-minded people became an organization.

They were just trying to protect their children, themselves, their pets, and their property and asked a paranormal investigator for help and to confirm they weren't crazy.

Unless someone is deliberately summoning paranormal entities with rituals, spells, curses, or by using a Ouija board, usually no one asks for a paranormal entity to disrupt the quality of their life. Most of us became interested in the paranormal by being involuntarily thrown into it.

For those who aren't reading this, I'm sorry that something showed up in your life that wreaked havoc. It's like cancer or an accident. I commend you for being brave. Picking up this book and taking a leap of faith, and trying your best to protect yourself and your community. To keep going when all heck is breaking loose is honorable and necessary.

Besides the victim, there are many kinds of people investigating the paranormal:

> **The ghost hunter (GH) -** goes to haunted sites and investigates and records any encounters with the paranormal. They post their results on social media and get a following, and may make money off of advertising, investigating, or selling their group t-shirts. They ghost hunt for mere entertainment and enjoyment and even perhaps a little narcissistic validation on social media. They typically investigate abandoned places. I value this type of investigator, and I know that ghost-hunting with your spouse, girlfriend, family, and neighbors is a great community occasion. It allows others to watch an investigation in a documentary or video without leaving their home. Their main goal is for entertainment only and unfortunately, their videos or documentaries don't show the GH helping others overcome or

remove their entity. We are useless as a profession if we can't figure out how to remove entities, or find other people or companies or services that remove entities. Our profession is just like being a firefighter. A firefighter shows up to answer a call from a person. If a firefighter and firetruck were to show up at a person's home and not put the fire out, the fire department would be useless and no longer in existence. The same is true for a paranormal investigator.

The Paranormal Investigator (PI) - goes to haunted sites and investigates and records any encounters with the paranormal. Then the investigator incorporates the sciences into the investigation to determine what makes the paranormal entity "tick." For example, they would compare the encounter with natural science, physical science, physics, chemistry, history, life sciences, earth science, geology, biology, biochemistry, microbiology, ecology, social sciences, anthropology, and history. The PI includes government people too. They may help law enforcement, mental health agencies, and others with information and resources. They are disciplined and structured and don't just run around aimlessly.

The Paranormal Agitator (PA) - This type of paranormal investigator investigates the paranormal with a disregard for public law and order and human decency. They trespass on private or public property or damage it intentionally. They shortchange their clients with services and fees. They don't deliver what they promise. Their unprofessional behavior results in legislatures and city governments' "cracking down" on the professional. They don't take care of their teammates or have first aid kits. It's mostly sensationalism and narcissism.

The non-believer - denies and discounts the paranormal. Demeans the profession. Presses moral indignation upon the profession. Everyone has experienced a "negative nanny." You know who they are.

The Cryptozoologist – studies Bigfoot, the Loch Ness Monster, Yeti, the Chupacabra, the Jersey Devil, or the Mokele-mbembe.

The Psychic/Medium – uses their gifts to study the paranormal and assist others. They may or may not reject matters of science. They are a valuable asset.

I'm not trying to be mean. We need all categories. The only choice we have is to pick a category we fit into and listen to the other categories to keep ourselves professional. I've learned more and enhanced my processes and services based on the feedback of a non-believer and an agitator more than once. For all I know by acknowledging the perspectives of skeptics, non-believers, and agitators, we open ourselves up to valuable feedback and opportunities for growth. It's essential to remain open-minded. Judgements are necessary for human survival, and no one wants to be judged harshly. As long as we can work together, and not bail on each other, we're all allowed to believe what we want to believe and think what we want to think. The important thing is to focus on the client, not our personality conflicts.

Ghost hunter

- Investigates sites

- Posts investigation video, photos & audio on social media.
- Obtains "likes' and popularity.

- Moves on to the next hunt

Paranormal Investigator

-Investigates sites

-Integrates science into the investigation (chemistry, biology, history, human physiology, medical science, mortuary sciences, electrical theory, etc.)

- Posts redacted investigation video, photos & audio on social media and protects the confidentiality and privacy rights of the client.

- Helps client remove violent or non violent entities.

- Shares research with other investigators, legislators, medical community professionals, and other government and private agencies to protect the living and their community from harmful or violent paranormal entities.

CONFLICT IN OUR OWN HOUSE

Just like any organization, there is conflict and people competing for ideas, resources, and attention.

Some of the areas of conflict within the investigation team and with the other stakeholders of the paranormal community include:

- Lack of consensus about the subject matter
- Limited availability of credible removal authorities
- Catholic Church believes they are the only and primary authority for paranormal removal
- Insufficient number of priests to meet the demand
- Strict standards for obtaining priestly services
- Lack of consensus and uncertainty in Interacting with paranormal entities:
- Lack of knowledge on appropriate interaction methods (best guess, try, and try something else)
- Measurement and interpretation of interactions
- Necessary resources and tools for effective interaction
- Communication with clients regarding the meaning of interactions
- Mental fitness of humans
- Ability to detect paranormal entities vs. ability to remove them
- Perceptions, stigma & efficacy of those who remove entities through spiritual means (mediums/psychics)
- Ensuring client safety
- Clients willingness to undergo removal efforts that go against their religious or other values
- Removal efforts may cause a paranormal situation to grow worse
- Client's ability to pay for investigation or removal services
- Shortage of investigation teams or removal services in small towns or rural areas
- Personality and intimate relationships conflicts with clients and team members, children

BIASES

We all have biases. We need them for the self-preservation of ourselves and society. I love our biases because they provide sanity checks for everyone and no one is exempt. I've learned more from a heated argument or discussion than from a book. We are professionals, not mothers, babysitters, attorneys, priests, or God almighty.

There are many types of biases:

Confirmation bias - people favor information that confirms or strengthens their beliefs or values.

The principle of replication - the experiment should be repeated more than once.

Illusionary correlation - perceiving a relationship between variables (typically people, events, or behaviors) even when no such relationship exists.

The Barnum effect - individuals give high accuracy ratings to descriptions of their personality that supposedly are tailored specifically to them, but are vague enough to apply to a wide range of people.

The power of suggestion - giving hints to obtain a specific outcome

The placebo effect- A benefit received from something that was not given (fake pill-but the pill worked as if it was a real pill).

The double-blind protocol- neither the participants nor the experimenters know who is receiving a particular treatment.

Occam's razor- a herd of running zebras sounds the same as a herd of running horses.

WHO WANTS TO ARGUE?

Because there is conflict in our house and there are no laws or standards for the profession, there are pros and cons with where we are as a profession.

We are unbridled and have free roam of abandoned buildings. But as we all know there are a few bad apples that ruin it for the rest of us. Enjoy the lack of legislation now, because it's not going to last.

How paranormal professionals present themselves and their investigation and investigation methods to the public today will influence legislatures in the future.

It's a double-edged sword, but things will be better. The internal squabbling between members of a paranormal team or squabbling between multiple paranormal investigation agencies will disappear because the legislatures will draw the line to some extent for the paranormal community.

We can assume that if someone is selling their house, they want to sell it quickly and at the greatest profit. We can assume that any haunting or violent acts about the house for sale will be watered down. I understand why. It is what it is. The old saying let the buyer beware includes the paranormal.

The reality is that government regulation of a profession means increases costs to the service provider. A bureaucracy in a company occurs when a legislative bureaucracy writes new laws.

Future legislation for our profession will most likely include the following:

1. Protecting consumers from predatory paranormal investigators and ghost hunters.
2. Property development decisions on haunted sites based on the violence level of the property (public park or a daycare center?)
3. Including paranormal intervention in mental/medical health and academics/EMS/Law enforcement.
4. Mandatory or voluntary reporting of haunted places and/or death reporting in real estate transactions (current and ongoing).
5. Standards for minimum levels of service (licensure).
6. Legislature or industry-approved remedies to remove violent entities.
7. LLCs, articles of incorporation
8. Mandatory minimum wage, workers comp, taxes, non-profit agency filings,
9. Methods of lawsuits for violations of the law, statutes of limitations,
10. Certifications and calibration standards of specialized paranormal investigation equipment
11. The paranormal investigation report is made a valid legal document in a criminal or legal proceeding, similar to how DNA testing is a lawful, valid test for paternity,
12. Public hearings and public inputs on these legislative agendas for those who support or don't support the laws,
13. How each state will legislate the profession in comparison to other states, and the reciprocity between the states.
14. A paranormal investigation report will be as common as a home inspection or appraisal.
15. Trespassing and access to public and private property.
16. Children and paranormal encounters and investigations (including mandatory reporting).

The way we professionalize something is that we have methods, tools, techniques, and outputs that are consistent with logic, sound judgment, competence, and the focus and awareness that we are to take care of our clients and therefore take care of our community. Every business has to have competence, judgment, planning, and a focus on serving their community. Not just follow a social media community with reckless abandon.

If we keep our house in order as paranormal investigators we can keep legislation to a minimum. A ghost hunter argues. A paranormal investigator investigates professionally:

- Inform our clients and the public when paranormal encounters are mistaken for naturally occurring phenomenon.
- Utilize specialized tools and techniques to identify and ward off harmful paranormal entities.
- Employ knowledge and expertise to safeguard individuals and communities from potential harm.
- Develop and implement effective strategies to minimize the negative impact of paranormal encounters.
- Familiarize ourselves with local, regional, and national regulations governing paranormal investigations.
- Ensure adherence to legal and ethical standards while conducting our investigations.
- Collaborate with authorities, legislatures and clergy and respect their jurisdictions and limitations
- Gather comprehensive data on paranormal phenomena through systematic investigations and data collection methods.
- Analyze collected data to identify patterns, trends, and potential correlations related to paranormal activities.
- Collaborate with researchers and scientific communities to contribute to the understanding of the paranormal for the betterment of society.
- Continuously enhance our knowledge and expertise in understanding how paranormal entities manifest and interact with the living.
- Develop and refine investigative techniques to improve our ability to gather evidence and analyze paranormal occurrences.
- Stay updated with advancements in relevant scientific fields to integrate scientific methodologies into our investigations.
- Educate and train our team members on paranormal investigation protocols, techniques, and safety measures.
- Disseminate accurate information to the public about paranormal phenomena, debunk misconceptions, and promote safety awareness.
- Prioritize the well-being and safety of team members and the public during investigations.
- Promote critical thinking in the profession, not just sensationalism.

WHY ARE THEY HERE?

Even a 5-year-old asks that, and so does a 90-year-old. Why is any of this here? The Earth, as it is, doesn't even have to be here either!

Our paranormal visitors are in this realm for many reasons. If we understand why they are here, it may help us figure out how to make them want to leave this realm.

Inability to Cross into the Light:

According to Mary Ann Winkowski, in her Book "When Ghosts Speak, Understanding the World of Earthbound Spirits, ISBN 13 978-0446581332 and ISBN 10 044658133X, some entities remain in this realm because they are unable to transition to the afterlife or "cross into the light." Other reasons include:

1. They were ordered to come or stay.
2. The entity had dementia or Alzheimer's at the time of death and might not have the cognitive ability to know they can cross into the Light.
3. They stay here because they like a person, place or thing here.
4. They don't want to leave loved ones behind (support, guilt, protection, revenge, apology, forgiveness).
5. They don't want to cross into the light for fear of punishment by God.
6. They don't want to cross into the light for fear of continued arguments with their ancestors if they fought while living.
7. They are here to protect the living from trauma they experience in a place the ghost lived at.
8. They are here to influence the timeline.
9. They have a specific message or purpose.
10. They appear just to prove to us mortals that they exist and have no other purpose.
11. They can't return (abducted, portal access).
12. The entity was summoned or remains in this realm by a Ouija board session or a séance session/medium, spell, or curse.
13. They wish to preserve the original aesthetic of a house and may mix up home renovation plans.
14. They can travel to and from realms at will.
15. The entity lacks the energy to leave.
16. They are here for asylum or punishment.
17. They are here to give us clues about their own society.
18. They are here for pain or pleasure.
19. Their presence is a result of a biological or chemical irritation or reaction

So, this begs the question about pet and animal ghosts. Our horses, dogs, and cats don't "live their best life," and they don't "have regrets." They also don't have a "bucket list" of things they would like to do or experience before they pass away.

We can assume that pets remain in this realm as ghosts because of incompetence and extreme loyalty to their owner or fellow pets.

THINGS AND COLORS

EVERY SINGLE TIME

In my extensive research and personal experience with paranormal investigations, I have observed recurring themes and environmental factors that seem to be associated with the presence of paranormal entities.

Dominant Colors:

One interesting observation is the prevalence of paranormal activity in places where the dominant colors are blue-green, teal, or light pink

Antiques and Dolls:

Another notable pattern is the frequent occurrence of paranormal activity in locations where antiques and dolls are plentiful. This includes Indian arrowheads and artifacts and urns of deceased animals or humans.

This doesn't mean that having these colors or objects in the home CAUSES paranormal encounters. It simply means that the paranormal prefers these colors and objects over any others.

The two ghosts that I had in my house appeared after I purchased an antique school desk and an antique trunk. As soon as I disposed of the antique desk and trunk, the paranormal activity stopped.

I believe that paranormal entities attach themselves to these objects because:

- They want to make sure it gets preserved after their death.
- The wood preservatives (formaldehyde) in the antique wood furniture act as a fuel source. Coincidentally formaldehyde is used at funeral homes and mortuaries, and medical centers to prevent a deceased human body from decomposing too quickly.

I believe that the colors in these haunted places are a soothing color to the paranormal and if given a chance they will choose locations with these objects or colors over any other.

In 90% of the initial interviews I had with my clients, they said they had these objects and colors.

I want my clients to remove a paranormal entity as quickly, cost-effectively, and safely as they can. I tell my paranormal clients to remove those colors and objects to make it less comfortable for the paranormal to want to stick around. Sometimes that's all it takes.

When my clients don't want to throw out grandma's 300-year-old desk, or the dolls or arrowheads or ashes and urns that have been in the family for generations. I ask them to put their antiques in storage or off the property and then see if the paranormal activity stops. If the paranormal activity stops when the antiques are gone, we know the paranormal entity was attached to those objects.

The client won't want to bring back all the antiques at a time once the paranormal activity stops. They would just be bringing the entity back into the home. They have to return one antique at a time, wait a few months and see if there is paranormal activity, then return another antique. If an antique item is returned and paranormal activity picks up or comes back, the client can assume that the paranormal entity is attached to that particular antique they just brought back.

One of my clients had dozens of pieces of antique furniture in her home because her mother was an antique furniture dealer. She refused to get rid of her antiques, which had a paranormal entity attached to them. She did end up moving her antiques to the basement. The entity remained. She even had a young child who knew there was a ghost in the house.

I was disappointed that I had to throw out (actually burn) $800 worth of antiques to get rid of the two entities I had in my home. But it was worth it. The hassles of having two entities in my home versus missing an antique desk or trunk were fair trade.

I firmly believe that when a person has a ghost, and they decide to move to another house, the reason why the ghost follows them to their new house is because the entity is attached to the dolls, antiques, ashes, and Indian artifacts. Literally, as someone buys or sells these items, the entity can travel anywhere on the planet Earth with them.

If people don't want Grandpa's antique desk to go to an undeserving buyer, then I ask that they donate the antiques to a museum.

I firmly believe that this is why so many museums and antique stores are haunted.

SLEEP, DREAMS AND THE PARANORMAL

So much confusion of the paranormal revolves around sleep.

Dreams are a natural biological function of all humans because dreams remove short term memory over the lifetime of a person. If I am 60 years old, I don't need to remember what time I had a bowl of cereal on the 4th Saturday after my 6th birthday. Dreams serve a crucial function for our emotional and psychological well-being. They allow us to process and make sense of our feelings, emotions, and experiences. By replaying recent events and incorporating them into dream narratives, our minds can actively work through unresolved issues and provide a space for emotional exploration. While it is true that humans are not biologically programmed to remember all their dreams, the act of dreaming itself plays a significant role in our cognitive and emotional development.

The misconception that humans MUST remember their dreams is a fallacy.

Hypnopompic hallucinations occur in the morning, as you're waking up. They refer to vivid sensory experiences that occur as a person wakes up from sleep. These hallucinations can involve visual, auditory, or tactile sensations, often presenting as dream-like scenarios. They can range from fleeting images or sounds to full-blown immersive experiences. Common examples include seeing shapes, figures, or objects in the room, hearing voices, or feeling a presence nearby.

On the other hand, hypnagogic hallucinations occur as individuals fall asleep.

Research suggests that a significant percentage of the population, up to 70%, may experience hypnagogic or hypnopompic hallucinations at least once in their lives. These phenomena are typically considered normal and do not indicate an underlying mental or physical illness. Sleep deprivation, irregular sleep schedules, and stress can also increase the likelihood of experiencing these hallucinations.
Up to 70% of people experience them at least once. They generally don't indicate an underlying mental or physical illness, though they may be more common in people with certain sleep disorders.

The astute paranormal investigator studies sleep hygiene and determines if a client's paranormal encounters are simply related to sleep hygiene issues.

There are hundreds of web pages that provide dream interpretation. I'll let those speak for themselves.

I personally remember seeing things, hearing things, and feeling things while sleeping throughout my life. I thought I was possessed or had a ghost or demon. Then I got diagnosed with sleep apnea and got a CPAP machine. Problems gone.

Here are some known medical conditions involving sleep that may cause a person to think they are having a paranormal encounter when it's only an issue with sleep hygiene:

Exploding Head Syndrome (EHS) is a sleep disorder. Although it sounds painful, it is not painful. With EHS, the person hears a loud noise or explosion in their head. The sound isn't real or heard by others. It happens as the person falls asleep or when waking up during the night. EHS is harmless and not a sign of another serious health condition. It usually doesn't require treatment.

Uncomfortable or powerful dreams that cause us to worry and think we are face-to-face with the paranormal can be caused by several factors:

- nicotine or nicotine patches,
- mashed potatoes,
- foods with starches,
- an uncomfortable mattress,
- allergies to food,
- laundry soap,
- perfume,
- candles,
- bedding,
- curtains,
- carpet,
- cleaning detergents,
- dust,
- fabric softener,
- fresh paint,
- incense,
- radon,
- carbon monoxide,
- methane gas (sewer gas) or
- new carpet glues.

Other reasons why people have trouble sleeping or dreams that appear paranormal include:

- pajamas that are too hot or too cold or too tight or uncomfortable.
- body heat changes as a person ages,
- the need to urinate or have a bowel movement,
- room temperature changes,
- pillow preferences versus availability,
- hip problems,
- back issues,
- neck/shoulder/waist positioning,
- pregnancy,
- excessive cell phone viewing,
- television watching and blue light exposure.

Other sleep issues include:

- Hearing loss,
- Hearing impairment,
- Sleep apnea,
- Background noise/white noise,
- Radio/tv/music needs change to get to sleep or stay asleep,
- Adjusting snooze bar duration or quantity, MSGs, foods with excessive chemicals and preservatives, and dehydration.

Faulty electrical wiring, outlets, switches, sockets, and electrical grounding can create a false perception of paranormal entities and sleep problems. I had one client that had wash cloths and socks floating and moving around because the electrical grounding and wiring in his home was so messed up that static electricity was everywhere. Electricity fixed, paranormal problem solved.

Even practical jokers can disturb a sleeper!

Other known medical conditions can impact our sleep:

- insomnia,
- hypersomnia,
- restless leg syndrome,
- narcolepsy,
- heart disease,
- lung disease,
- snoring,
- late night snacking,
- shift work,
- circadian rhythm changes,
- jet lag,
- eye masks,
- sleepwalking,
- bed wetting,
- eye problems,
- sinus problems,
- toothaches,
- eardrum disorders,
- pacemakers,
- oxygen machines and
- medications/supplements

What to do? Get a physical, eye, ear, vision, heart, sleep apnea, tests, MRI, and CAT scan, and make sure the sleeping issue isn't a known medical condition. Assume it's paranormal or empathic after you're medically cleared or are getting treatment for known medical conditions. Keep yourself and your family safe. Join social media groups and read books that discuss your sleep issues.

I ask many clients to video record themselves sleeping. They may see something at the foot of their bed or other entities in their home. They don't have to watch every video the next day. They can just watch their sleep videos if paranormal activity occurs.

The first thing I tell my clients is to figure out their sleep hygiene. If known medical conditions or other conditions all check out then they can assume that their situation is paranormal.

SLEEP PARALYSIS

Many paranormal clients complain of having "sleep paralysis," which further makes the person paranoid that they are having a paranormal experience.

However, there are two kinds of "Sleep Paralysis":

> **"Medical sleep paralysis"** occurs each time human dreams to prevent the human body from moving while the dreamer moves in the dream. The human body has mechanisms to prevent a human from moving its arms or kicking its legs if the person is asleep and is moving and kicking their arms in a dream. This prevents the human from falling out of bed or actually acting out their dreams.
>
> **"Paranormal sleep paralysis"** - In the paranormal realm, a shadow person or other paranormal entity may press down on the human chest, or touch the human, or the bed or give the human other sensations that the entity is in contact with the living human.
>
> These two types of sleep paralysis are confused as being the same.

Mary Ann Winkowski, in her book "When Ghosts Speak" stated that our deceased family and friends can speak to us in our dreams if the deceased person has crossed into the "Light". She also believed that it's easier for our deceased family and friends to talk to use in our dreams than it is for a ghost to try and interact with the living while we are awake (a haunting). Her lifelong profession confirms that yes, the paranormal speak to the living in our dreams.

THE OUTDOOR GHOST

What about the client that calls the paranormal investigator in a panic because they saw something on their outdoor home security cameras or outdoor entity?

1. Paranormal entities outdoors may be **"passing by"** and may not have an interest in a home or property.
2. If the paranormal entity in the outdoor camera has an interest in the property and plans on sticking around, then the house will exhibit other signs and symptoms of a paranormal entity taking up long-term or permanent residence.

When I have a client that sees something from their outdoor camera? I tell them to clean off their camera with WD40 to prevent bugs and cobwebs from sticking to the camera lens. The lens only has to be cleaned about every six months.

Yes, bugs and cobwebs will appear on the camera lens in the winter. Then I tell clients to check back in a few months to determine if the entity was just passing by or not.

It's extremely difficult to track and monitor a paranormal entity outdoors. The current modern electronic **"gadgets"** a PI uses don't usually work at long distances.

WHAT SMELLS?

When a client speaks about smelling things that are out of the ordinary they can appreciate that there are known medical conditions that can be mistaken for a paranormal encounter:

Phantosmia is a known medical condition where a person can smell things from the past. For example, smelling smoke from a deceased relative who used to smoke, love flowers or perfume from a deceased relative who was fond of those flowers or perfume. It's common among Veterans, firefighters, law enforcement and emergency service personnel. This condition is also associated with dementia or other mental health disorders.

"Parosmia" is different from phantosmia. Parosmia is smelling something that smells like something else. For example, you might smell chicken but are actually sniffing toothpaste. If a person smells a dirty baby diaper this is associated with those who have brain tumors. The smell of rotting eggs or dead things is an indication of an evil entity.

As always, it's a good idea to check with your doctor or an ear, nose, or throat specialist. If there is no known medical reason for your symptoms assume it's paranormal and go from there.

HOW DO WE DEFINE “HAUNTED”?

Part of being a professional investigator and not just a ghost hunter is that the paranormal community as a whole must decide and have a standard as to how we define “haunted”. Even if the definition is just in a dictionary. A paranormal investigation team isn’t taking a dictionary to an investigation site.

Our clients are making major life decisions such as moving out of their haunted house, or they are protecting their children or pets so it’s critical that the final investigation report tells the customer, clearly, how “haunted” is defined.

There are two types of hauntings:

1. **Residual** – the entity does not know it’s there. For example, a ghost ascends or descends a flight of stairs at the same time or day because there was a traumatic event involving the stairway.

2. **Intelligent** – The ghost can interact with the living. Deliberate. This would include poltergeist or demons. Poltergeist simply means “noisy ghost”.

People are going to believe what they want to believe, and everyone doesn’t always agree on what is or isn’t. That’s OK as long as the client knows what both sides of the issue are, and the client can make their own decisions.

So, for practical purposes, this is how **“haunted”** and paranormal are defined in this book.

Paranormal - something that can’t be logically explained using current methods of science, common knowledge, education or wisdom.

Haunted:

1. People and witnesses have seen/heard/felt/witnessed things firsthand (ghosts, orbs, things moving, knocks, scratches, doors slamming, etc.)
2. Electronic gadgets confirm something (paranormal gadgets, lights flicker, batteries drain, etc.)
3. Pets respond to things that aren’t there.
4. There are things in the home and colors in the home that are consistent with things and colors in other places that have been declared haunted (dolls, antiques, teal or pink items)

THE CONFIDENTIAL REPORT

As professionals, we do our best to provide our clients with relevant, factual information that is scientific, and the data is coherent, realistic, and can easily be incorporated into the religious beliefs or value systems of our clients.

They get our best advice and research when we prepare and submit a final investigation report to them.

The pre, during, and post-investigation attempts to rule out known scientific explanations, naturally occurring phenomenon, lens glare, practical jokes, and medical conditions that are mistaken for a paranormal encounter.

Once known medical, scientific, or natural reasons are filtered out, then it's paranormal or a haunting.

Our clients don't want to invest thousands of dollars in electronic paranormal gadgets or spend hundreds of hours researching their paranormal problems. They don't want to spend thousands of dollars on a medium or psychic, or other removal agency, especially if the removal service seems "kooky."

They most positively don't want to see investigators arguing about whatever is or isn't in front of them or their children.

They most positively don't want to see two paranormal investigation agencies at the same site arguing with each other, either.

Our purpose is to bring peace to the client.

Argue in private.
Praise in public.

Many paranormal teams have a medium or psychic or specific method for paranormal removal they bring to the table. Most do not. Hopefully, we have told the client that removing an entity is not like calling a plumber who can fix a problem with a part, 90 minutes of labor, and then the bill is paid, and all is well. We know the removal of entities can take days, weeks, months, or years.

The final investigation report tells the client what the gadgets discovered, what conclusions were made, how pets responded, what happened and to who and when and clearly states where there was disagreement in the profession or between investigators. It gives the client the opportunity to make their own decisions based on their beliefs, values and particular situation.

The final report allows the client to decide to keep working on removal efforts, let a docile paranormal entity remain, and, most importantly, allow the client to make a major life decision-***move out*** of the house or stay in the house.

Our client has a certain level of economic capacity, physical and emotional energy, and a certain level of social support in their family and communities. Our answers in our final report give our client a game plan for a short or long-term solution to their paranormal problem. Our paranormal report answers the questions the client asked before or during the investigation.

The investigation team might not be able to get a list of all the violent acts that happened in the past at an apartment complex they are investigating, but the homeowner, as homeowner, is able to get more information because they are the actual owner.

My client knows I'm going to remove their personal identifying information from their report and share my research data with paranormal stakeholders. They agreed to this arrangement before I start the investigation. Suppose they don't want me to share the report, I won't. Most clients want me to share the report with the public. The client's life has subtly been interrupted, and they will gladly share their research so others don't have to go through the same frustrating situation.

Here is a sample of a final investigation report. The report is customized to suit the situation, client, or level of expertise on the investigation team. This is what the client needs and expects. Chances are, by the time the client has called you, all options are on the table, and since most people aren't religious or churchgoers, they will most likely contact a paranormal investigator before they contact a member of the clergy as the very last resort:

PARANORMAL INVESTIGATION REPORT

FOR

NAME REDACTED

ADDRESS REDACTED

LINCOLN, NE

April 3, 2023

The purpose of this paranormal investigation was to:

Determine if the home

a. is "haunted".
b. is a hazard or threat to humans, animals, the environment, natural resources, or Native American interests.
c. causes "imminent danger" to humans or pets or damage to property as a result of paranormal activity.
d. if humans may be possessed by an evil entity.
e. provide the homeowner with information to make decisions in their best interest.
f. to validate that the homeowner is not "going crazy".

FUNDING:

This paranormal investigation was conducted at no cost to the customer as a community service by Lincoln Paranormal Interventions. The paranormal investigation team is not certified to remove entities.

CAVEATS

The paranormal investigation doesn't subscribe to any specific, religious or political agency. The investigation attempts to provide known scientific research and uses research from other confirmed paranormal sites and experts. The client can obtain additional investigations. Whatever decisions the client makes as a result of this report would align with the economic, political, social or religious belief system of the client. The paranormal investigation team will never recommend a course of action that would deliberately cause harm or death to humans, damage to property or disruptions to civil society. The client is aware that they must address the pros and cons of any course of action versus the risks to rewards of a course of action. The investigation won't provide any séances, Ouija board sessions or other rituals or rites other than a listening session. Rarely, if ever do cultic rituals or other forms of divination lead to long term happiness. Initially they provide excitement and entertainment, and in the long-term result in a negative lifestyle.

THE INVESTIGATION TEAM:

Kirk
M****
J****

GENERAL CONSENSUS AS TO THE PARANORMAL INTERACTIONS:

The resident reported seeing balls of light (orbs) and hearing voices, whispers, conversations, or being told what to do, and hearing arguments.

BASIS AND RATIONALE FOR THE INVESTIGATION:

There aren't any laws or statutes in Nebraska that a real estate agent, home buyer, home seller, or landlord disclose to anyone if any paranormal activity currently or previously occurs on a property. The homeowner felt that a paranormal investigation would provide additional helpful, relevant information to make other decisions. The homeowner wants to make sure the occupants of the home are safe.

The Paranormal "Body of Knowledge" (BOK)

There is no "official" agency that certifies if a place is haunted. The collective knowledge of all paranormal research is open to interpretation and argument as to what is considered a paranormal encounter and what isn't. However, most paranormal sites exhibit certain similar characteristics among people, places, pets, and technology. Where there is a dispute among paranormal researchers, the dispute is provided to the client, and the client can determine what they wish to believe or not.

The customer was provided a checklist of items that typically occur at a home or place where a paranormal activity occurs. The list also contained items that are known medical, naturally recurring, or scientific phenomena that can be mistaken for a paranormal encounter. The checklist was used to rule out known, explainable phenomena and what to focus on during the paranormal investigation that might be paranormal.

The homeowner's checklist answers indicate small amounts of paranormal activity consistent with the Paranormal BOK.

To determine if a paranormal entity is present in a home the following factors are considered:

1, The human 5 senses of the homeowner, guests, occupants, and investigators
2. The behavior of pets
3. Readings from electronic devices
4. Environment or activity in the home that occurs with no reasonable explanation
5. Confirmed research obtained from the home occupants, neighbors, private and, public records
6. Objects in the home that are typically found in homes where paranormal entities are present
7. Stress and health of the occupants of the home in dealing with the unexplained

If the people, pets, property, environment, and technology all indicate signs and symptoms of a paranormal entity, then it's assumed the property is haunted.

The homeowner has no pets.

The investigation team arrived at the home. They signed waivers and confidentiality agreements with the client, set up equipment, established lines of communication, ground rules, safety, and "buddy" systems, and discussed paranormal encounters in the home. The group was told that there are no "taboo" topics and every little detail matters.

"Triggers' ' were set on the kitchen counter to elicit responses from a paranormal entity. The "triggers" were samples of Isobutate, sunscreen, baking soda, corn starch, powdered sugar, yeast, flour, sugar, saline solution, salt, acetone, hydrogen peroxide, glue, isopropyl alcohol, boron, magnesium, teal paint, CBD and film developer. These triggers are chemicals that interact with the environment that cause objects to appear or act as catalysts for the production of energy sources.
The team investigated the basement, first and second floors, and the outdoor areas of the home.

The SLS camera was used outdoors and didn't detect any entities.

Next door and across the street neighbors weren't interviewed.

Equipment used: Dowsing rods on wood frame, black and red Rook detector, 2 black motion sensors with antenna & alarm, gray EMF detector, black pistol digital thermometer, red pistol thermal camera, black dosimeter (gamma ray tester), red laser grid, spirit box, black pistol sound amplifier, windmill, an SLS camera, and a baseball. The dosimeter was used to check for white colors and the EMF detector was used to detect black shadow figures based on the ROYGBIV rainbow spectrum of light:

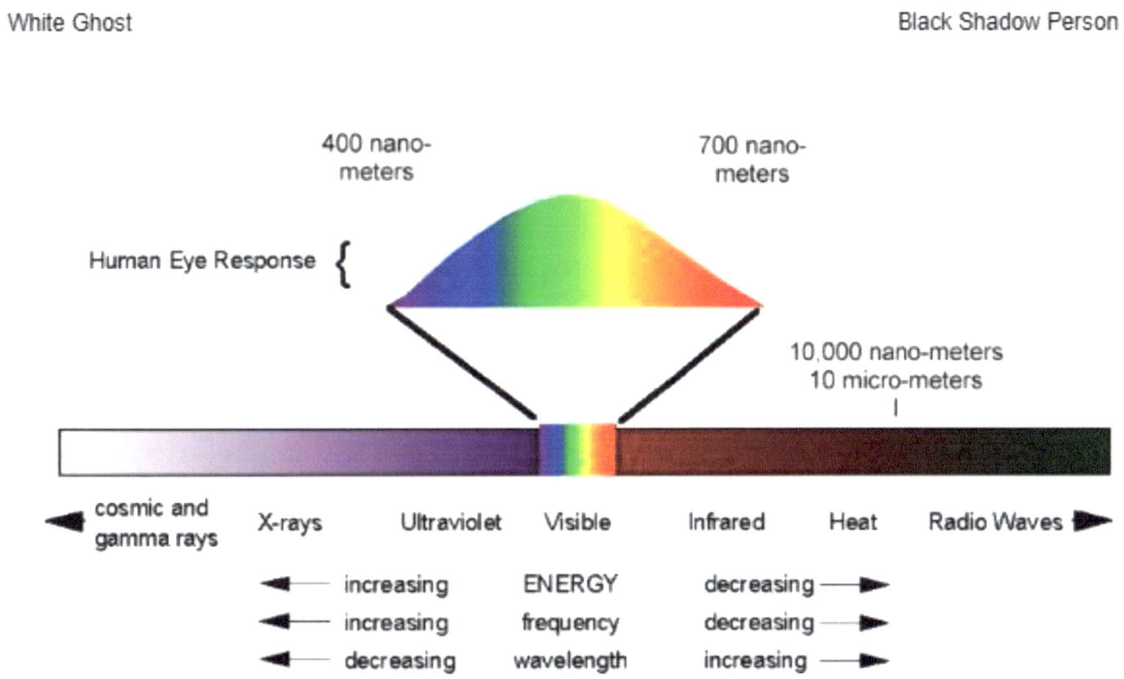

Image source: https://socratic.org/

No equipment was available to test for:

Delta wave - Deep sleep 1-4 Hertz

Theta wave - 4-7 Hertz Meditation/sleep

Alpha wave 7-12 Hertz - not coherent and relaxing

Beta wave - 12-40 Hertz - Conscious reasoning

Gamma wave - 40-70 Hertz - Self-aware, happy, learning state

The occupants of the home were told how the equipment operated, what information the equipment provided, and how the information collected was related to the paranormal. The customer and investigators took turns using the equipment. The client was asked to unplug the microwave oven at least an hour before the investigation team arrived to eliminate or reduce any false microwave readings.

The team held a 30-minute "listening session" in the master bedroom and another 10-minute listening session in the garage next to the basement storage room. The trigger devices were the baseball, motion sensor, dowsing rods, and SLS camera.

One person on the team asked the following questions to elicit responses from entities: The team asked any entity to answer by touching a trigger device once for "yes", or twice for "no".
Are you female? Are you a follower of Satan? Were you ordered or forced to stay in this realm? Are you able to cross over to the light? Do you want to cross over to the light? Can I help you cross into the light? Can I do something to give you peace so that you can cross into the light? Were you sick before you passed away? Did you commit suicide? Were you murdered? Was there an accident? Do you fear punishment by God if you cross over into the light? Do you fear arguments with others already in the light? Are you here just to prove to the world that you exist? Do you have a message for someone? Were you summoned here by someone else? Are you here because of a curse? Can you travel to this realm at will? Are you here for a specific person? Are you here because you like a specific object or thing? Are you here because you are being punished? Are you here for asylum? Are you here to protect someone? Are you here to apologize?

To determine if an entity was or wasn't present 4 laws and principles of knowledge were applied:

The law of non-contradiction, the law of causality, the basic (and imperfect) reliability of our perception and senses and the analogical use of language.

INVESTIGATION EVENTS

One team member walked throughout the house with the black pistol audio amplifier but didn't hear anything out of the ordinary.

One team member scanned the home with the dosimeter and found no changes in radiation waves.

The EMF detector had high readings explained later in this report.

The homeowner used an SLS phone app and detected a stick figure in the master closet and a smaller-sized stick figure on the wall. The phone app results were compared with the actual SLS camera which confirmed the entity.

MASTER BEDROOM:

During the listening session:

1. The motion sensor went off when the first 3 or 4 questions were asked.
2. The homeowner reported his left foot was touched and got cold during questioning
3. One investigator said the corner of the black blanket on the master bedroom floor flopped by itself.
4. One investigator reported seeing the baseball move away from the bedroom towards the kitchen.
5. One investigator saw the stick figure from the SLS camera on the right side of the master bed and then the stick figure moved onto the homeowner. The stick figure remained there consistently.

6. Two of the three team members heard footsteps walking from the kitchen through the breezeway to the lower-level stairs.

LIVING ROOM:

1. The living room motion sensor tripped incessantly so the battery was replaced and the sensor worked as normal.
2. There was a black colored plastic dragon statue on a living room shelf that held on to an SLS stick figure.

BASEMENT:

1. Several stick figures appeared on the SLS in the basement garage. At least two in total.
2. When the team asked if the entity had been dead "...for at least 16 years…" all three investigators heard a female voice the spirit box say "yes' ', and the motion sensors flashed and beeped.
3. Even after moving furniture and household items around the garage to confirm if the stick figures were a false positive the stick figures maintained their presence.
4. Furnace room EMF readings were 0 until the meter was pointed at a wall and peaked at 198 mG. It is believed to be an energy distortion or portal because the homeowner said there were no EMF appliances on the other side of the wall.

EMF meter pointing at the basement wall

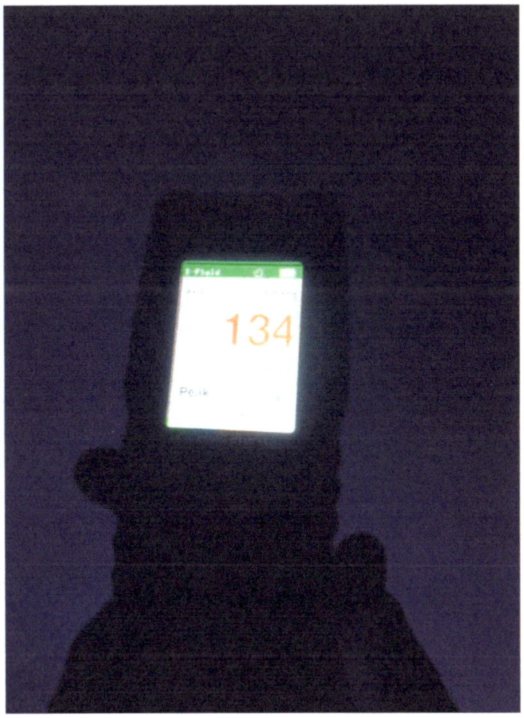

Throughout the investigation, the stick figure was docile, calm, and very attached to the homeowner.

The homeowner believed it might be a deceased person in the family.

The general consensus is that the home contained more than one entity, and based on the paranormal activity and electronic sensors, it was a docile female.

DAMAGE TO PROPERTY OR INJURIES TO PEOPLE:

Post investigation: NO property was damaged, and no personal injuries or death was reported after the investigation.

POST INVESTIGATION:

14. The day after the investigation the homeowner reported an object fell off of the clothing dryer and no one pushed the object off the dryer. Things in the home were generally quiet.

Yellow colored figure is a living human
with corresponding SLS stick figure.
The teal color shape and corresponding SLS stick figure is not a human being.

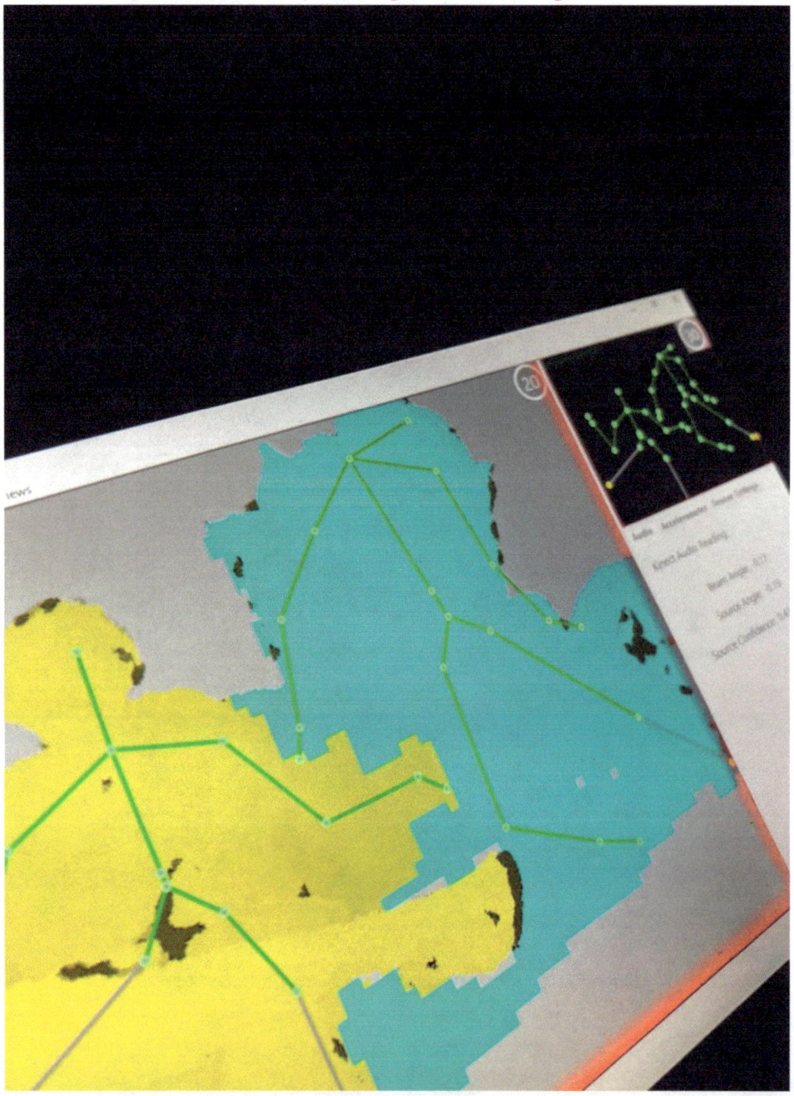

SLS stick figure
Sitting on wood bench in living room

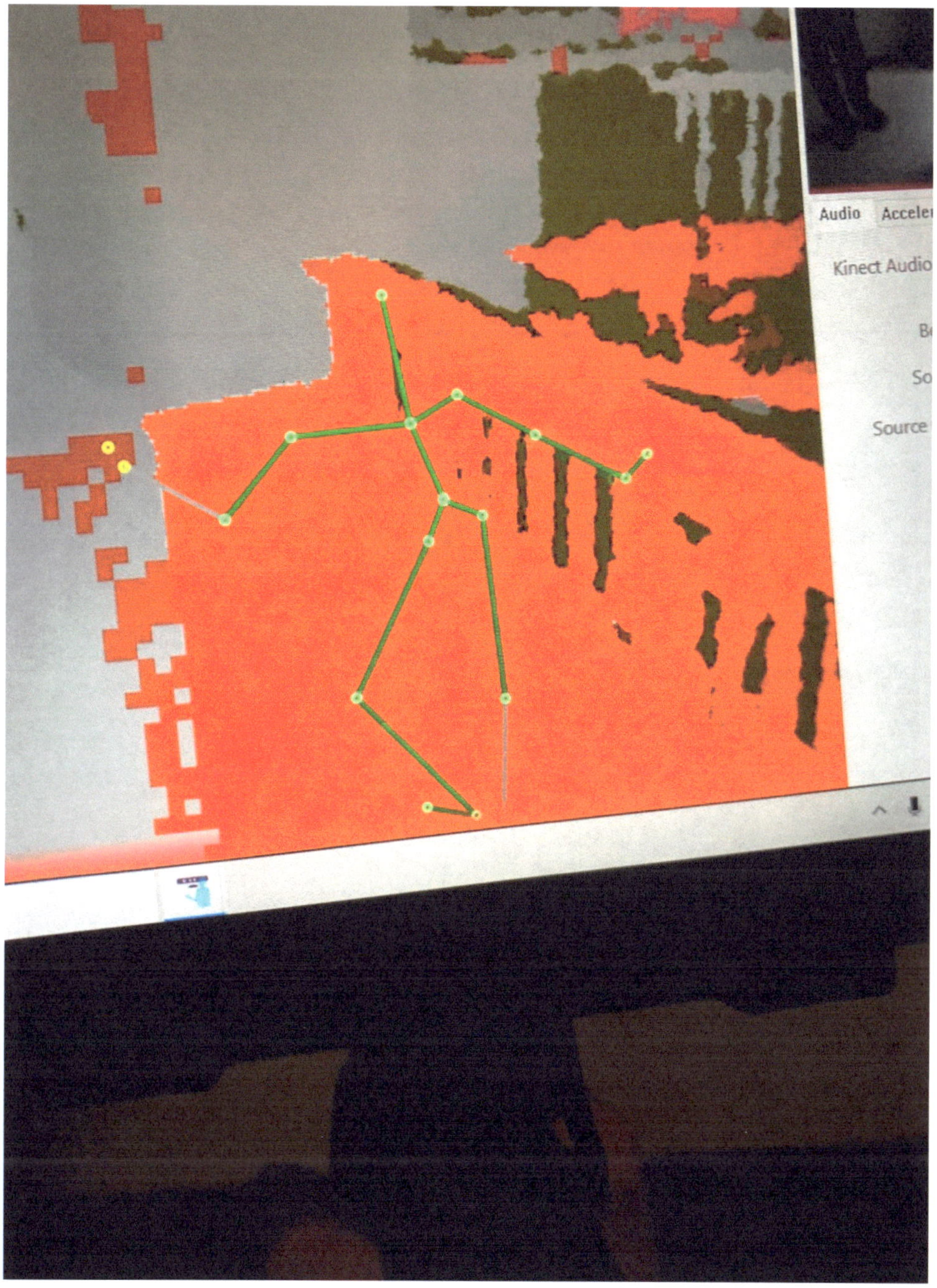
Audio
Acceler
Kinect Audio
Source

SLS stick figure in basement

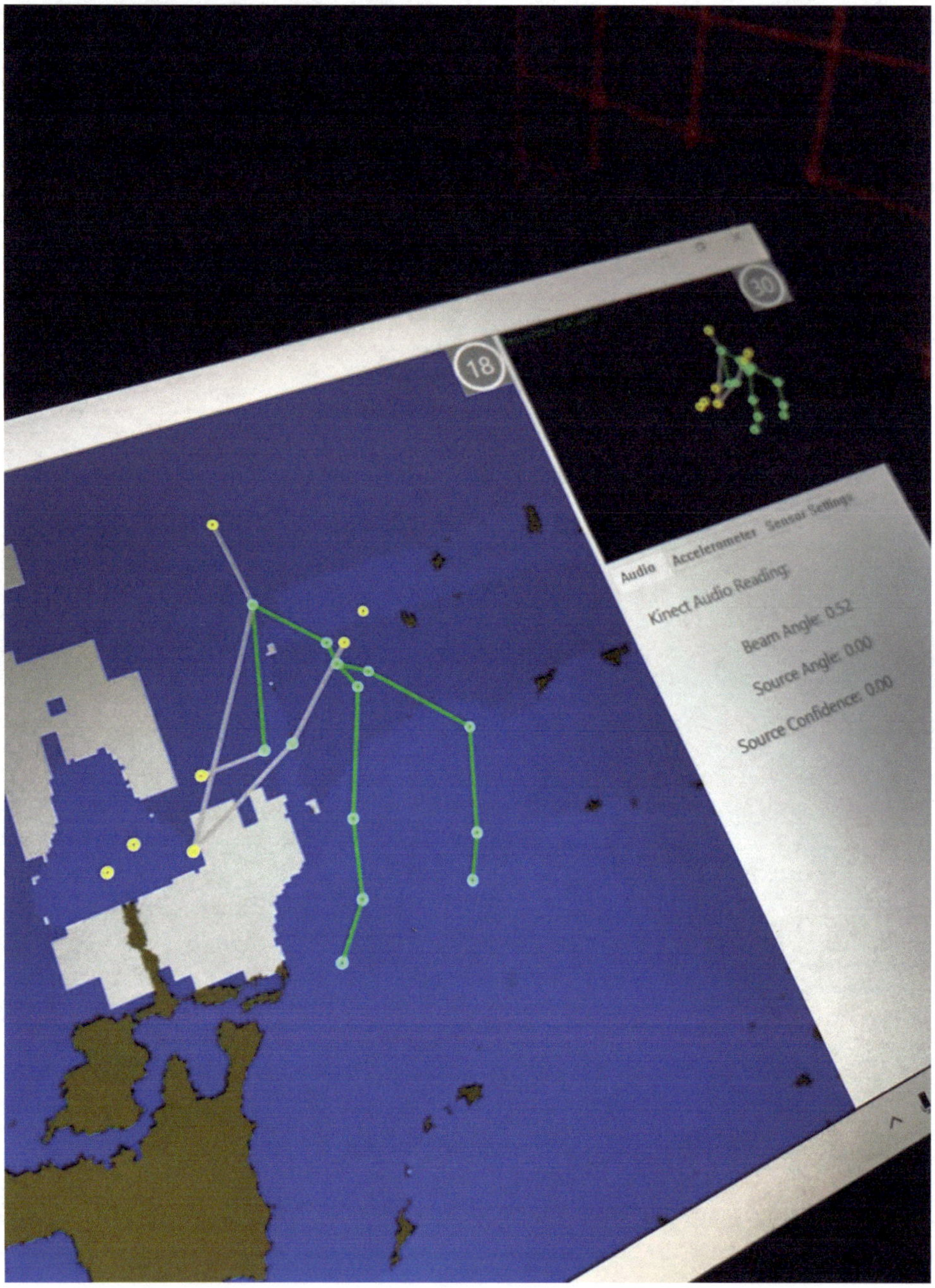

CONCLUSIONS/ANALYSIS

The client reported that their skin felt chilled. This is consistent with the research that blood is negatively charged because of the heparin in it. A paranormal entity would require some type of energy to manifest, and the negatively charged energy from the human blood would be extracted from the human, causing the skin to chill from the loss of heat and energy.

Source: https://www.vedantu.com/question-answer/blood-is-a-charged-sol-a-positively-b-neutral-c-class-12-chemistry-cbse-5f8cf79946566e310be95a66

IS THE HOUSE HAUNTED?

Yes.

Cambridge dictionary defines haunted as "... A haunted place is one where ghosts often appear."
Merriam-Webster dictionary defines haunted as" to visit or inhabit as a ghost.
The Free Dictionary defines haunted as: "…To inhabit, visit, or appear in the form of a ghost or other supernatural being."

The residents and investigators experienced unexplained phenomena in the home that can't be explained by known science or natural order.

A digital" haunting" occurs when a paranormal entity is detected through an electronic device but is not detected through the normal human five senses.

Because of the excessive number of stick figures captured on the SLS camera, indoors and outdoors, there may be more than one entity in the home.

IS THE HOME A HAZARD OR THREAT TO HUMANS, ANIMALS, THE ENVIRONMENT, NATURAL RESOURCES, OR NATIVE AMERICAN INTERESTS?

HUMANS:

No. The entities didn't appear violent or aggressive.

ANIMALS:

No. The homeowner had no pets in the home.

IS THE ENVIRONMENT OF THE HOME CAUSING PARANORMAL ACTIVITY?

Possible. The home may contain objects that paranormal entities are attached to.

DOES THE HOME IMPACT NATIVE AMERICAN INTERESTS?

Unknown. The City of Lincoln is most closely associated with the Plains, Great Sioux and Lakota tribes. The investigation team has limited access to the property records of the homeowner and therefore isn't able to draw any conclusions as to Native American interests. It's not known if the house is on sacred Indian or burial grounds. As a homeowner they can contact agencies that deal specifically with Native American culture, rituals and their society:

Nebraska Commission On Indian Affairs
indianaffairs.state.ne.us
1445 K St Ste 6, Lincoln, NE 68508
(402) 471-3475

United States Dept of Interior
Bureau of Indian Affairs
Indian Affairs
1849 C Street NW
Washington, DC 20240

University of Nebraska-Lincoln
Native American Indian Studies
Ted Hibbeler
Office: 402-444-7804
Cell: 402-906-0439
ted.hibbeler@unl.edu

UNO Indian Studies
College of Arts and Sciences
222 University Dr E, Omaha, NE 68182

ARE THE OCCUPANTS IN "IMMINENT DANGER" BECAUSE OF THE PARANORMAL ACTIVITY?

No. Not at the time of this investigation. Some entities become friends with home occupants and then turn evil. The entities appear docile and calm. But these entities seem calm and passive.

ARE THE HUMAN OCCUPANTS OF THE HOME POSSESSED BY AN EVIL ENTITY?

Not at this time. The homeowner was asked to continue monitoring household members for signs of possession. At the time of the investigation no humans displayed signs or symptoms of demon possession. Stress and anxiety levels of the home occupants were consistent with a typical family that has been dealing with unexplained activity. The family exhibited consistent signs of both anxiety and relief when they saw their first stick figure on the SLS camera that is consistent with other people who experience SLS imagery for the first time.

Signs of possession include: smelling rotten eggs or flesh, abrupt changes in menstrual cycle, superhuman strength, unnatural body movements, self-mutilation or suicide attempts or thoughts, divination abilities, loss of appetite, becomes a medium, starts to discuss unbiblical or cult-like topics, stops blinking their eyes regularly, eyes change color, victim of sexual abuse or sexual trauma, acting in ways that don't suit a particular personality, use drugs or alcohol to cope with pain, severely depressed or feeling hopeless, anorexia, etc., ancestors committed murder, suicide or sexual molestation, voices in a person's head to tell a person to do immoral or harmful acts, a person asked Satan to exchange the human soul for something, multiple phobias or fears, abandoned by parents, disabling sessions of depression or rejection, failed repeatedly at significant relationships, serious health or financial issues, hindered in bible reading, prayer or church services, loss of hearing or speech functions, or becoming unstable.

RECOMMENDATIONS

1. If the customer wishes to continue investigating "self-help" they can rent paranormal electronic equipment. The deposit is $50 and each item rental is $20 for a 3-day rental. A rental contract will be signed for each device.

2. The client may wish to get a full medical physical, hearing and sight test and mental health checkup with sleep apnea study to rule out known medical conditions that may be mistaken as paranormal encounters.

3. The entity may be a former resident of the home. Discovering the actual name of the entity may facilitate productive interactions or the removal of it. The homeowner was provided a public record of the former occupants and owners of the home that may aid in determining the name of the entity.

4. The investigation team doesn't have the people or resources to remove an entity. Here are some suggested low cost methods to remove an entity. If the homeowner desires to hire a third party for removal it's recommended the homeowner pay 50% up front and the remaining 50% of the removal efforts once the entity is gone for at least 6 months. Then the homeowner won't get ripped off.

5. If the customer believes that a ghost is attached to a person, the person can go to a funeral home and ask the attached entity to cross into the "light" with the other deceased person at the funeral home. Details of this process are in the book "When Ghosts Speak" by Mary Ann Winkowski. ISBN: 0446581186, ISBN 13: 9780446581189.

6. House cleansing. A suggested house cleansing method is attached to this report from the book "How to Clear Your Home of Ghosts and Spirits: Tips and Techniques from a Professional Ghost Hunter", by Debi Chestnut. ISBN 10 0738739316

7. Talk to trustworthy friends and family about your experience to ensure your safety and to retain credible witnesses.

8. When performing removal actions do so with more than one person. Pets may be negatively influenced by removal activities. Removal attempts may cause more paranormal activity.

9. Remove objects in and outside the home that paranormal entities are typically attached to such as dolls, antiques, Indian arrowheads, and sun catchers. This includes jewelry, furniture, rugs, vases, candlesticks etc. It also includes items purchased at garage/yard sales, second-hand or thrift stores, gifts from others or estate sales.

10. Objects can be removed for the short term. If the paranormal activity decreases or stops once the objects are removed, then move one object at a time back into the home every several months to determine what specific object the entity is attached to and then discard the object.

11. Remove saved ashes of deceased relatives or pets or place them in smaller containers.

12. Break mirrors before disposing of them.

13. When disposing of an item that is believed to have a paranormal entity attached to it render the object useless, and to ensure the object won't be taken out of the trash can and simply moved somewhere else.

14. Statistically, paranormal entities prefer the relaxing colors of pink, teal, light blue, and blue-green. Repainting walls of these colors may be an incentive for the entity to leave. Painting walls with these colors doesn't CAUSE paranormal activity to occur. It simply means that if given a choice, paranormal entities prefer these colors over others.

15. If a suggested removal activity doesn't work, learn from it and move to something or someone else and keep trying.

16. If the customer wants their house blessed or to discuss demons or possession they can contact the Catholic Diocese of REDACTED at REDACTED, NE at 402-REDACTED 488-0921. The Church may require a person to undergo a mental health evaluation before taking on exorcism procedures.

17. If applicable, the homeowner would hold one more séance, ritual or Ouija board session and ask all entities to leave, not attach then destroy and dispose of the Ouija board and any and all items used for rituals.

18. Never, ever, ever, ever ghost hunt or perform a paranormal investigation or house cleansing alone.

19. Get a Bible, Crucifix, Rosaries or whatever protection items your faith calls for.. You can get holy water from a Catholic Church if you bring your own empty water bottles. Paranormal entities will not go near organic salt. Make salt circles if necessary. You can also pour holy water on your hands.

20. If lights are flickering, TV or computer screens are sputtering, and batteries are draining or not charging, have a licensed electrician confirm your electrical wiring, outlets, switches, appliances, and devices are properly grounded and fixed and work according to the manufacturer's instructions.

21. Video record, sound record your experience. Record at more than one angle. Get a witness. You may need audio, video, or a witness if you need to hire a paranormal investigator, hire a medium, psychic, house cleanser, or to help get exorcism approvals from the Catholic or Christian church or for medical attention at a hospital if the entity is violent.

22. Report any paranormal investigation results to your community leaders, businesses, real estate agencies, historical societies, and others with a vested interest in keeping your community safe and developing your community for growth using the investigation results to guide the decisions. A heavily haunted area or an area that has a significant amount of paranormal activity would be developed for businesses first, child care centers, schools, churches, and residential areas last. Develop the area incrementally to ensure paranormal entities don't scare away residents or businesses. Once a house or community gets the label and stigma of being "haunted" the quality of life of the community typically deteriorates.

23. Sit down with your family and talk about a strategy on how the ghost will be discussed in the home, not at home, and when guests or family visit the home, schoolmates, co-workers, daycare providers, pool cleaners, housekeepers, etc.

24. Although not common, some entities have departed a home when a bell was rung in each corner of the house.

25. Placing salt at your front door and around your windows is also known to remove the paranormal entity.

26. Confirm from your local or state environmental office if your paranormal experiences are a result of environmental reasons. (rocks, springs, wells, pits, caves, underground rivers, seismic activity, minerals and elements in rocks, etc.)

27. Burning sage like a candle doesn't remove a ghost or paranormal entity. The whole house must be saged and cleared. Here is a book on how to cleanse your house with sage: "How to Clear Your Home of Ghosts & Spirits: Tips & Techniques from a Professional Ghost Hunter" -by Debie Chestnut ISBN 10 # 0738739316.

28. Ask the entity if you can help it with its purpose or message or to return to the light or to achieve what it came for as long as humans aren't harmed and property isn't damaged and you are able and willing to help.

29. Treat the entity the way you want to be treated including saying please and thank you.

30. Tell the entity you are friendly and want to communicate.

31. You can talk to the medical community and scientific community about your paranormal experience. This ends with mixed results. Typically, governments and law enforcement won't assist to resolve negative or harmful encounters between humans and the paranormal unless there is personal injury, death, property damage, or an imminent threat to public safety, property, or infrastructure or until it's too late and something bad happens. Scientific and medical communities typically dismiss the paranormal as those institutions are more focused on the physical than the spiritual or metaphysical.

32. If you or a member of your family develops superhuman strength, unnatural body movements, engages in self-mutilation or suicide attempts, has divination abilities, becomes a medium, starts to discuss unbiblical or cult like topics, starts to lose hearing or speech functions, or becomes unstable then chances are the person is possessed by a demon and a Catholic Priest should be called for an exorcism consultation.

33. If the entity is a "renovation ghost", cooperate with it for legitimate reasons. A renovation ghost attempts to retain the original aesthetic of the home and may tear down shelving, hide tools, hide project materials or disrupt renovation efforts. It typically doesn't bother the homeowner once renovations are complete.

34. If the entity is a purposeful ghost or messenger ghost it will depart when its message is delivered and UNDERSTOOD, or when its purpose is finished.

35. Ask the entity questions that are easy to answer. (Knock, scratch, pull, push).

36. Set up boundaries with the entity as appropriate.

37. Homeowners who are childless found that purchasing child toys for child ghosts have been successful and beneficial to the entity/human relationship.

ADDITIONAL RESEARCH AND INVESTIGATION

As the non-homeowner, the investigation team has limited authority to access public and private records on the history of the property. However, the homeowner has better access. These civic and government agencies may be able to assist the homeowner with additional information as to the history of the home. The client was provided with a letter template to send to these agencies to research more information about the property.

Nebraska State Patrol4600 Innovation Drive Lincoln, NE 68521
Lancaster County, Nebraska Sheriff Department 575 S. 10th Street Lincoln, Nebraska 68508
City of Lincoln, Nebraska Police Department 575 S 10th St Lincoln, NE 68508
Federal Bureau of Investigation Lincoln NE Field Office 100 Centennial Mall North, 363 Lincoln, Nebraska, 68508
History Nebraska Headquarters 1500 R Street Lincoln, NE 68508
Lincoln Lancaster County Genealogy Society P.O. Box 30055 Lincoln, NE 68503
Lancaster County Register of Deeds 555 S 10th St Ste 102, Lincoln, NE 68508

>>>>>> THE LAST RESORT<<<<<<<

After multiple removal attempts when is it time to leave and vacate your house and move?

1. The entity causes physical violence to humans or animals or damage to property.
2. The safety and health of the children has reached a dysfunctional level.
3. The family is no longer comfortable spending more money on removal attempts.
4. Arguments between home occupants have resulted in long-term negative consequences.

SPECIAL NOTICE

If a paranormal entity is attached to objects in the home, then the entity moves to the new home with those attached objects.

THE REMOVAL OF THE PARANORMAL ENTITY

The paranormal investigation team assists the client, clergy and other professionals to help removing paranormal entities, especially violent ones. The client has many factors to consider, which would be considered when preparing their final report. These factors include:

-Lowest cost and requires the least effort to permanently remove the entity.
-The removal effort doesn't cause the entity to be more violent or more active
-The removal effort doesn't go against their spiritual, religious, or moral beliefs
-They may have limited options for third party removal services if they are geographically separated from a major metropolitan city where there are fewer resources.
-That the psychic, medium, or other ritual or ceremony doesn't add more paranormal complications as a result of the removal efforts.
-They can keep their household goods and preferred colors and decorations in their homes.
-They only need and want one removal effort, one time.
-That if they decide to remove the entity "self-help," they do the self-help activity correctly and are properly trained on how to perform the self-help activity and have the necessary self-help tools to do it.

Where each person draws the line on how much they want to help the client is up to each investigator.

I draw the line when a client asks me for money for things that aren't related to a paranormal removal.

I had a client that wanted me to buy her college textbooks, clothing and new couch because her "entity" was preventing her from achieving a lifestyle she wanted. I did offer to help my client pay for sage and other paranormal removal items. But I'm not a sugar daddy or an ATM. I had another client who asked me for money to move out of her apartment because she thought her boyfriend was possessed by a demon. I try to help who I can, when I can with what I have, but some people will take advantage of others if given the chance. Decide up front in the final investigation report what is acceptable to ask for and what's not.

If you take away nothing else from this book other than this then the world will be a better place:

If someone hires someone else to remove a paranormal entity, get a contract. Pay 50% up front and the remaining 50% only if the entity is gone for at least 6 months. Then no one gets ripped off.

HEARING WHAT THEY WANT TO HEAR

I wish that at this point in human history that there was a 100% assured way to remove an entity. There's not and I feel horrible for that. I know that there are people who are victims of the paranormal and they didn't ask for a paranormal entity to disrupt their life.

I have the utmost respect for any human being that can remove an entity quickly.

The investigation report tells the client ways to remove an entity, that both the investigation team and client are financially and morally willing to undertake.

Some removal efforts take more than one try.

I'm perfectly clear about clients who hear what they want to hear, instead of listening to what I am saying.

I'm perfectly clear if I can remove an entity or not.

If a client asks for help from anyone and the answer is "no," the client thanks the person who declined and then moves on to ask someone else.

If the removal effort doesn't work after several attempts, ask someone else for help. This includes using social media, the Internet, or other means, even if the help is out of state or out of the country.

It is what it is!

Right now, in the human race, there are more investigators and ghost hunters than there are removal experts.

Right now, in the human race, there are more patients requiring medication than there are pharmacists to fill the prescriptions. Right now, in the human race, there is a shortage of 18-wheel truck drivers to deliver goods. The paranormal profession is subject to the laws of supply and demand, just like every other profession.

LAST RESORT: Move out of the residence.

I served two separate clients who didn't or couldn't have children, therefore no grandchildren either. Both of them decided to continue interacting with their child ghosts. The clients continued to buy toys for their ghost. Another client would place a ball at the top of the stairs and let the child ghost push the ball so it would go down the stairs. I get it. I would never do it, but I understand the need to connect.

COUNTY RECORDS & THE OLD FOLKS

I love people who are old. To some extent, their life was harder and in other ways easier. Lifetimes of experience and mistakes. I regret not spending more time with my grandparents talking about the past. Their version of history is much better and more personal than the history textbooks at school.

If my client allows me, I will go to the old lady next door or the two brothers who are aged 84 across the street and live their final years together as their spouses have already passed away. I'll ask them if they know of any paranormal activity around my client's house or the neighborhood.

The elderly love to share their stories and reminisce about the past and their youth. They will share personal accounts of paranormal encounters that they have experienced in their life and in their neighborhood.

The elderly WANT TO pass their wisdom on to the next generation, so the next generation can learn from their mistakes early on and have a better quality of life. They are begging to do it.

What a treasure it is for all of us.

I can even go to the county historical society and the library and look at the newspaper archives to see if there was any paranormal activity or violent activity at the home I'm investigating. (Hint, hint).

How much is a cup of coffee worth talking to a senior citizen? Once the senior citizen passes away, so does the story.

The homeowner or the property owner can get legal, criminal and historical information about the property better than I can. Simply because I'm not the property owner, and they are.

Some police departments will provide a criminal history record of violent activity at a property, with names kept private if needed. State, county, and local historical societies and libraries will do the same.

WHAT ABOUT THE CHILDREN?

I get asked this question all the time, and many of my peers disagree with the answer, which I am OK with. I have children, and I'm a paranormal investigator. This hits home with me as well.

There are things I tell my clients when it comes to the paranormal:

1. I am not the biological or custodial parent of the children at the investigation site. I am a mandatory reporter, just like everyone else. I am responsible for interacting with a child in a safe, respectable, professional manner that complies with the law and the values and norms of the parents or guardians.
2. I am not God, nor is anyone else.
3. A parent is responsible for the safety and care of their children in their home.
4. A minor child doesn't have the ability or wherewithal to live someplace else if they are scared or being tormented by a paranormal entity.
5. A child will resent their caregivers if the child is being harmed or tormented by anyone in the house, and the adult members of the house shrug it off and do nothing.
6. A child will remember, for the rest of their lives when their caregivers took visible steps to remove a paranormal entity and make a place safe, even if the caregiver thinks that the removal effort is crazy or won't work.
7. Child protective services, law enforcement, civil rights groups, judges, teachers, religious leaders, social justice warriors, neighbors, school principals, community organizations, courts, and the media will gladly tell the public if a child has bruises, cuts, injuries caused by anything, including a paranormal entity. The parents have to prove the paranormal entity did it and the caregiver didn't.
8. A child who has a home where they don't feel safe will be a child who is socially and mentally incompetent out of the home for years. This is true even if there is no paranormal activity where the child lives or not.
9. When I investigate a home with children I am a stranger walking into an environment and poking around in closets, basements, attics, under beds, cabinets and sheds, garages, and wherever the client says I can have access to. I must carefully explain why I am doing what I am doing if the parents don't.
10. Sit down with children in the home and come up with a plan on how the paranormal entity will be discussed and dealt with:
    a. At the child's school or other public gatherings.
    b. When people come over, so if the forks start sliding off the table without anyone touching them fork, everyone knows what to tell the guests.

Once the child starts talking about ghosts at school and has bruises or scars on their body, the school nurse, school mental health office, the bus driver, the playground supervisor, the cafeteria lady, the soccer coach, and the principal will follow their safety protocols and procedures as **"mandatory reporters."**

This is why a final, formal paranormal investigation report is so critical to the success of our families and the family unit. Taking a paranormal investigation report to the school will smooth things out with the school and simplify life.

When I get asked if a child should participate in an investigation, I tell the parent they are responsible for the safety and well-being of their children. They know the emotional maturity of their children better than I do. The parents know what values and morals they want to impart to their children.

If the parents decide that their child is going to participate in the investigation, then the same safety and investigation protocols I expect my team members to follow applies to the children.

When I provide information to the parents, the parents decide if and when, and how the child will receive the investigation information.

Everyone on my team signs confidentiality waivers and safety protocol agreements and waivers that are legally able to. Even if the child is in the other room watching TV while the investigation is ongoing.

WHAT ABOUT THE NEIGHBORS?

Unless it's a decided-upon lifestyle, no one wants to live in a haunted house, especially if the ghosts are wreaking havoc on the physical and mental conditions of those living there.

Unless the neighbors already tell each other, no one tells their neighbors they think their house is haunted. The neighbors may or may not tell a new owner or occupant the house was haunted before they moved in. The neighbors may have shrugged off the last people that occupied the home as being nut jobs. Every neighbor must decide if their actions at what level they will take part in their next-door neighbor's lives.

If my client is worried about the neighbors and a reputation, then when the client hires a paranormal investigator ask the investigator to:

A. Wear plain clothes,
B. Come driving unmarked cars,
C. Come after dark,
D. Come when the house is empty such as vacations or work schedules allow,
E. Only talk to the neighbors if the client gives permission.
F. Tell the neighbors the team is there for a "safety inspection". Which is totally true.

Paranormal conversations with neighbors are initially subtle and indirect. "Did you see the TV show about haunted houses last week?"

A paranormal entity taking up residence in a neighbor's home may be the reason why your home is getting paranormal activity also.

PARANORMAL RENTALS INC.

The paranormal investigator can loan the paranormal gadgets to the client, for a deposit and fee. Then the property owner can investigate when their children are sleeping so the kids won't get scared and neighbors won't snoop.

The investigator trains the client how the gadgets work, and what the reading means, how to change the batteries and those necessary things.

Here is a sample of an equipment rental agreement contract:

XYZ
Paranormal Interventions
3911 Smith Ave
Anytown, State 12345
123-456-789
email@email.com

Date:___________________

TEMPORARY ISSUE RECEIPT AND RENTAL AGREEMENT

My signature below indicates that I understand that the equipment I am borrowing is on loan and is the property of _________________. I understand and agree to return the items listed below in working condition. I agree to set up financial and other arrangements to repair or replace these items to a serviceable condition if I cause any damage to the equipment. I understand that when this equipment was issued to me, I received training on how to use it, and it worked when it was given to me. I must arrange with the issuer to return the property to the investigator. I understand that legal action may occur if this agreement is breached. The fee to rent this equipment is $_______ with a $______ deposit. My deposit will be refunded no later than 7 calendar days after the equipment is returned. If I damage or destroy equipment or lose it I will lose my deposit.

Issued to:

________________________________ ________________________________
Print Sign

Issued by:

________________________________ ________________________________
Print Sign

Date scheduled to return items to paranormal team: ______________________________

Items:

Page 1 of ___________

INJURIES, ACCIDENTS, THEFTS, DAMAGES?

An unexpired first aid kit is considered part of the paranormal investigation team's standard equipment inventory.

Do we have to conduct a pre-survey to make sure the investigation area is safe?

Who is responsible for calling EMS if there is an accident at an investigation site?

Who is bringing the first aid kit? What's in the kit? What needs to be in it?

What is the weather going to be when we arrive and for the duration we are there?

Who is responsible for collecting and gathering the research equipment when the investigation team has to leave the site when there is an emergency?

What are the safety policies and procedures at the investigation site or do they have to be developed because there are none?

Who decides to end an investigation based on the severity or number of injuries or accidents?

Who pays the ambulance bill, medical bill, police response fee for an incident?

Who contacts the police if something is stolen or an unlawful act is carried out at an investigation site, illegal substances are found and the like?

Who is trained in first aid? CPR, child first aid and child CPR and AED?

Is the buddy system used in case someone happens to someone?

Is there cell phone service at the investigation site?

How far away is the fire department, law enforcement, ambulance service & medical facility care?

Does the site have to be pre-inspected before the investigation to size up the location, equipment siting, power requirements and safety conditions?

Are the electrical devices, power panels, generators, mechanical switches and other devices that turn on or off?

Are things accessible to the paranormal team if the team must interact with machinery?

Are there hazardous materials at the investigation site? Wildlife, plants and trees that cause allergies?

Do we need to bring chairs/tables to keep track of our people and equipment?

Floors safe, structure is safe?

WAIVERS, HOLD HARMLESS

Since we are professional paranormal investigators we are clear about our expectations, assumptions and perceptions as to being safe and who is responsible for what if someone is injured or something is damaged. We provide waivers to our clients and our fellow investigators that clearly state who is responsible for what and who is responsible for whom. The waivers apply to the clients and each individual investigator on the team.

Here is a sample of a waiver:

WAIVER, HOLD HARMLESS, CONFIDENTIALITY

This paranormal investigation is provided as a public service, and the customer will not be charged for the investigation or for one (1) final summary investigation report. The customer will be charged only if they decide to have the team sage. Saging is optional. Payment shall be in the amount of $100, payable only after the paranormal entity is gone for at least six months. The customer agrees it has the authority to allow the paranormal investigation team to conduct an investigation on the property and will provide access to the property to conduct an investigation.

Should the investigation team damage any private property or the customer damages the property of the paranormal investigators, both parties agree to set up repair/replacement arrangements for the damaged property to restore the property to its original condition. Both parties agree to make arrangements to return any property inadvertently left behind on the property by the investigation team and return any property that the paranormal investigation team inadvertently removed from the property under non-malevolent circumstances.

The customer understands that the paranormal investigation team is responsible for obtaining their own medical care in perpetuity should they be injured on the property during the investigation and that the customer shall not be held liable for any medical bills, expenses, etc., related to this investigation. This includes injury caused by any paranormal entity, human error or defects or equipment, tools, unsafe areas on the site, etc. In the event of the death of any party, each party assumes the burial and other expenses separate from the other. The customer is responsible for the personal safety of their children, pets, and other visitors who are not part of the paranormal investigation team.

In consideration of this service, the customer understands and agrees to waive, release and forever discharge the paranormal investigation team, its subsidiaries and affiliates, and their respective agents, employees, officers, directors, shareholders, successors, and assigns from any and all claims as a result of any material or human damages caused by the paranormal entities on the property.

The customer understands that this investigation is done at the individual risk of each party and recognizes that the behavior of paranormal entities is not always predictable and may be dangerous or fatal. The customer understands that any attempt to remove paranormal entities from the property does not always work 100% of the time, and then paranormal entities may become violent and create more paranormal activity during any removal process. If the customer decides to allow the paranormal entities to remain on the property, they do so at their own risk, expense, or peril, absent the investigation team. The customer is under no obligation to accept the recommendation of the investigation team. The customer is aware that simply investigating a paranormal entity may cause more activity to occur before, during, or after the investigation.

The Parties agree and acknowledge that the existence and the terms of this Agreement and any oral or written information exchanged between the Parties in connection with the preparation and performance of this Agreement is confidential information. Each Party shall maintain the confidentiality of all such confidential information, and without obtaining the written consent of the other Party, it shall not disclose any relevant confidential information to any third parties, except for the information that: (a) is or will be in the public domain (other than through the receiving Party's unauthorized disclosure); (b) is under the obligation to be disclosed pursuant to the applicable laws or regulations, rules or orders of the court or other government authorities; or (c) is required to be disclosed by any Party provided that such information and the parties involved shall be bound by the confidentiality obligations similar to those set forth in this agreement. This stipulation shall survive the termination of this Agreement for any reason. This includes any medical information.

An investigation report that contains personally identifiable information about the customer is the sole property of the customer to be used as the customer sees fit. An investigation report redacted and not contain any personally identifiable information about the customer is the sole property of the investigator to be used as the investigator sees fit.

The customer understands that the investigation methods and information is current in the paranormal body of knowledge at the time of the investigation. The spirit and intent of the investigation are to provide the customer with actionable information for his or her own purposes, safety, and quality of life. The investigation report doesn't deliberately attempt to advocate for a particular religion, ideology or political belief, product, or service. The report and services are provided "as is" with no warranties other than this agreement.

By placing my signature below, I acknowledge and agree that I have read this agreement. I understand all of the terms and conditions contained herein and that this agreement will be in full force and effect during the initial investigation, pre-site survey, investigation, and post-investigation.

Date: ____________________

Client & Paranormal Investigator(s):

Print names: Sign names:

TWISTED HELP

I spent some time with two separate paranormal clients, several months apart. Neither knew the other, and didn't live near the other. Both had a shadow person occupying their life and lifestyle. Both contacted me to validate their experiences and wanted a third party to confirm they weren't losing their marbles. The photos they showed me were clearly shadow people.

A shadow person is a paranormal entity that appears like a human shadow. However, the shadow person has no face, no fingers or toes, but has arms and legs and hands. A hat-wearing shadow person is the leader of a hierarchy of a group of shadow figures who don't wear hats. Some shadow people wear bow ties, and some have colored eyes. I'm not sure why the shadow person wears the fedora hat from the 1920s era or bow ties of any period in human history as symbols of their hierarchy, but it is what it is. The bow tie would indicate another vertical or horizontal hierarchy in the shadow realm. A bow tie would indicate a male entity, a fedora hat a male entity, but the fedora hat was worn by females before males wore them as common apparel. Some cultures and religions believe that if a human passed away and was possessed by a demon or one of Satan's followers that the human automatically became a shadow person at the time of their passing.

The environment in their home and the behavior of their pets were consistent with a paranormal entity occupying a home. Both of the clients were concerned about the time and energy they were investing in the entity and how much time and energy the entity was investing in my clients. I could sense and hear the anxiety and fear when I interacted with them.

But…. both of them claimed and showed me photographs of them engaging in sexual activities with their shadow person. The mirror in one client's room was the portal for the shadow person to enter this realm. A hat-wearing shadow person instructed three non-hat-wearing shadow figures how to properly have sexual relations with the living female. Reminiscent of the story in the Bible where fallen angels left heaven, came to Earth, and started relations with human females.

Neither client asked me how to remove the shadow figures from their lives. Both said the experiences were the best sexual encounters they ever had. When I asked both of them what they told their family, friends, and co-workers, their reason for not interacting with the living, and their excessive amount of alone time with the shadow people, neither of them knew what to say. Neither of them wanted to tell their parents, siblings, or others. Neither had a long-term plan for the special shadow that was now an intimate part of their life.

It bothered me that I had to tell both of these clients that I didn't want to investigate them and I turned them away. I could have spent months researching the shadow figures firsthand and could have obtained vast amounts of information about the shadow realm.

I don't usually spend time investigating paranormal entities if my client isn't suffering some kind of pain or torment from the entity. The longer I got into the heads and minds of the shadow figure, the more time they got into my mind. Where is a safe tradeoff?

Both were madly in love with their shadow figures. They were promised high positions in the "shadow hierarchy" if they committed suicide and joined the hierarchy.

I asked each of them to write a book about their experiences. I wished them well. I hated to do it.

For those of us who are living, we will all be deceased and will have all the time in the universe to interact with those who have passed away before us when we pass away-for eternity. The quality of the relationships that these two clients had with fellow human beings must have been poor enough for them to choose a shadow person for a lover, friend and companion.

Every time I meet a client that is dealing with a shadow person, including now, as I write this book my computer gadgets go haywire. I've had computer monitors burn up and as I type these words right now the word processor is typing words that are misspelled or typing extra letters or none or putting spaces where there are none. My spell checker underlines the misspelled words with red lines but there are no words.

It's part of the risks of the game with this profession. Sometimes a fire department lets a burning building burn down because whatever is in the building is unsafe or it's just too risky to send firefighters into the building. So, it is with the paranormal.

The Fedora Hat of a Shadow Person
Source: Wikipedia.org

WHAT OF THE BIBLE, THE CHURCH, THE CLERGY & SCIENCE?

I believe that the church and clergy want to help their communities. I have no problem with the Church. I also believe that science and the Church are not as far apart as most people might think.

No other book in the history of humanity has caused more division or brought more people together than the Bible. I'm going to offend someone in this chapter but I won't be offended if you skip this chapter. I would prefer if you read it because preventing an argument before it even starts is how people become successful.

It's actually easier to read, interpret and understand the Bible than it is to study the paranormal. Any person who studies both will tell you so.

Jesus healed the sick, removed demons, saved souls, and performed miracles for humans who weren't Christian or Catholic. This is not the case with the modern Catholic Church when it comes to removing demons or other paranormal entities. It is what it is.

Members of clergy don't always agree with us, and some members of the clergy don't walk the talk. They are human just like us.

We live in a society where God is a convenience item and people pick and choose when, where, how and how much they want religion in their life.

Our client and fellow investigators have their own religious and spiritual beliefs.

Many of my clients are church going Christians or they have religious beliefs, many are not.

The reality is a paranormal entity picks and chooses who they will interact with, regardless of the religious beliefs the person has.

My clients are going to believe what they want, say what they want and do what they want and steer their offspring in the same direction. So are you, so am I.

How does the paranormal investigator incorporate religion into the investigation without arguments and division?

When I investigate, I specifically tell my clients that I don't endorse a particular religious doctrine or belief. I do share religious topics with my clients as they bring them up. They must bring them up first.

Before I investigate, I ask the client what their dos and don'ts are for a paranormal investigation and I try to incorporate their religious and spiritual beliefs in the final report and in our investigation. If they tell me not to go into this room because there is exposed asbestos in the room, and they tell me what is taboo and what isn't.

Usually the room they don't want me to enter is the one I need to investigate!

I tell them our investigation is confidential and nothing is taboo, and that every single thing that happens or is seen or heard in an investigation is relevant. I open the door for honest dialogue. If we can't talk about difficult ideas, we shouldn't be in the business of dealing with difficult paranormal entities.

I let the Bible speak for itself. I don't add or remove Bible verses or words. There is a reason why a topic is in the Bible and there is a reason why topics aren't in the Bible.

A cookbook and the Bible both have a purpose and an intent as to why it was written. I never met the authors of the Bible or the author of the cookbook I have on my bookshelf. I didn't write the cookbook or the Bible and wasn't there when the cook made the recipe and then wrote the recipe into a book. I wasn't around when Jesus lived. None of us were there either. The playing field is equal. None of us were there, none of us wrote it, and 2000 years later, we still argue about it.

Since we weren't there and didn't write it, we can believe the authors or not.

Those who have arguments with the Bible can argue with the authors of the Bible who wrote it and not argue with the people who just read it.

In a majority of the societies on Earth, no one is forced to read the Bible as a condition of employment. No one is forced to read a Bible to buy a car. No one is forced to read the Bible to eat food or buy groceries. Bibles are not in most public schools, and memorizing Bible verses or writing papers on Biblical topics is not a requirement to graduate from school. No one has to read the Bible or quote from it to get a car loan or home mortgage or operate a lawn mower or blender. There is a separation of church and state in most places. Most members of the clergy are not breaking down the front doors of our homes at 2 am demanding we read a Bible or we will lose our house, pay fines, and be jailed. (I said most-not all).

We all know, or at least we think we know, that the Church (Catholic Church especially) has or might have information on ghosts and the supernatural. As a human race, we have been asking the Church for centuries to reveal those secrets to humanity. They haven't. They won't. That is the reality. But the more they conceal and the longer they conceal it just means problems for those who conceal later on. No one gets away with everything forever, even if they get away with it now. No one. Obviously, those who conceal believe that the risk to share is worse than the risk not to share.

I am grateful for science in that we can study the universe in an organized, comprehensive, coherent manner. Science has provided much, good and bad.

The paranormal is a profession where science and the supernatural meet. A clinically dead person shows up and disappears at will as a ghost, and we bring scientifically manufactured devices to where the dead person showed up and we study it, then post it on our scientifically manufactured cell phones and social media sites.

Centuries ago, "scientists" said the Earth was the center of the universe. The Church agreed and said God created the Earth in the center of the universe. Years later, "scientists" said that the Sun is the center of the universe. The church agreed and said God created the Sun as the center of the universe.

Years ago, "scientists" said human females carry a baby in their womb and the church agreed and said human females carry a baby in their womb. Years later "scientists" said the male carries the sperm and females carry the egg and the sperm and egg combine and make a baby. The church agreed.

These are the same "scientists" who say they know the internal temperature of the planet Mars, but have never been to Mars and stuck a thermometer into the core of the planet Mars to take the temperature. We give scientists credit where credit is due, and we give them a little breathing room when we can. Mars is pretty far away and Mar's isn't going to be offended if we tell someone the incorrect temperature of its inner core.

Science and the church have always disagreed, for thousands of years, and we are all still here occupying the same Earth just fine. Most members of the clergy are not breaking into high school science classrooms at night to destroy the frogs stored in formaldehyde that are being studied in the name of science. (I said most.) Most science teachers are not demanding members of clergy use science textbooks for Sunday sermons. (I said most)

Thankfully, the political and verbal bashing between political groups and the Church is louder than the verbal bashing between members of the science and the Church. All have dug their trenches to fight the fight.

The Catholic Church, as the oldest and most well-known provider of demon removers is, by Canon rules, required to have at least one paranormal expert in each diocese. The expert may get some training and how much experience the expert gets depends on how much paranormal activity the priest is asked to intervene in.

Under Canon 489 Section 1 each Catholic diocese is required to have a locked file cabinet with confidential information that is limited to public access and even limited access to what is in the locked cabinet to members of the church itself.

Other religious denominations have paranormal experts in their institutions. In 1999 the Catholic Church opened up its doors for non-Catholic members of the clergy to attend exorcism training classes that were previously only exclusive to Catholic priests. The Catholic Church also released its exorcism manual to the public.

Thankfully, many Christian members of the clergy who aren't Catholic are excellent paranormal investigators and successful exorcists. Those who can't obtain exorcisms through the Catholic Church can go elsewhere, and some have been successful. Not everyone can meet the high standards the Catholic Church imposes on their paranormal clients.

If the Church standards for humans to receive paranormal services and exorcisms (free or not) are extremely high, and the number of exorcists is few, then other Christian denominations and the private sector have no other option than to equip society with exorcists. People are sometimes getting violently harmed by paranormal entities, whoever helps first and solves the problem first wins.

We are overloaded with paranormal investigators and ghost hunters; we need more exorcists and more people who can quickly remove violent entities.

I share my paranormal investigation reports with the Church. We all have to work together. I can assume that the Church knows things that I don't know about the paranormal and that I know things that the Church doesn't know about the paranormal. I won't turn down help, nor will a church. I know local churches won't spend money on electronic gadgets.

The church may require someone to undergo a psychiatric evaluation before the church expends a lot of resources into an exorcism or removal of an entity.

I had a client who called the Catholic Church to ask for help with a paranormal entity. The church went to their front door, gave my client some religious pamphlets, and told my client to pray and "get right with Jesus," and then left and never came back. I asked that priest if he wanted to join me in the investigation, but I found out he labeled me as "nothing but trouble".

I had to write letters directly to the Vatican for another client because all the priests in her geographical area wouldn't help my client. The letter worked, and within a few months, the Vatican dispatched help.

I had another client who had a priest, Catholic exorcist, Catholic Bishop, psychologist, and the paranormal team all working together on the same case. It happens all the time, and it works. As I have said in this book, the client takes advice from ALL the experts (church, gadgets, scientists, mental health professionals) and makes their own decisions. Just like the priest is the intermediary to remove a demon with spiritual gifts, power and authority, the paranormal investigator is just the intermediary to remove a ghost with gadgets and research.

Support from the church varies.

Bottom line:

The church responds to "assertive" requests but does not usually respond to panic or pressure or threats. When asking the church for help don't be afraid to go to the top to get the help you need. The Catholic Church is a bureaucracy, so we must speak to the bureaucracy in a bureaucratic method. The Catholic Church is full of males, and males are logical beings in the human species. We must communicate with the Catholic Church in a logical manner. I know this to be true:

- No one can call up the church and get help the same day.
- Most likely a person wanting paranormal help from the church will have to pass a mental health test.
- Document the paranormal encounters on paper, go as far back as you can, don't skip anything – a local church may have to go to a bishop or the Vatican for permission to perform paranormal removals. The Vatican responds to paperwork and photos.

Many of my clients don't want to do paperwork. I tell them that when they journal and document their paranormal encounters they are actually healing when the get it all on paper.

Share your research with clergy, agree to help the client, and not argue about ideology. Millions of people and millions of businesses, and millions of institutions have solved problems throughout human history with or without arguments.

The framework of religion, clergy, science and the Bible that I present to my paranormal clients is:

1. The Bible was never intended to resolve ALL matters of science to perfection.
2. The Bible clearly tells humanity that demons and the paranormal exist, that we are to avoid them, and Jesus Christ removes them from the human condition.
3. The Church follows science and only studies science to achieve the primary objective of the Church, which is to bring souls to Jesus Christ as a path to eternal life.
4. The Church will help to the extent they are able and willing. If the church, in any denomination, provides help, we have to play by the rules of the church to obtain and retain the help.
5. The Church believes God created "science" and everything else.
6. Agree to disagree, share research, and cooperate.
7. If the church or scientific communities can't or don't want to remove harmful demons and violent paranormal entities from the lives of the human race, we have a moral obligation to protect our fellow humans from violent entities absent either or both.
8. Members of clergy and scientists are humans, treat them like humans.

9. Members of the clergy and the scientific community are in their positions and most likely must adhere to their institutional ideologies or they will be removed from their positions.
10. Let the client decide what is best. Keep the focus on the client. All we can do is provide the client with resources and information. We can't force a client to accept it or take our advice or help.
11. Scientists exist for the sake of science; the primary role of the Church is to bring people to Jesus Christ. Everyone else is optional.

I have asked the Catholic Church what specific details or information they need from a paranormal investigation team in order for the Church to begin removal or exorcism efforts. They only said to contact them and that trained clergy will handle it. I still give them my reports and cooperate. I'm sure they know things that I don't and I know things they don't. I'm happy with their answer. At least we are communicating!

POSSESSION

Possession signs and symptoms from the Catholic Conference of Bishops and multiple web searches:

1. Abrupt changes in the menstrual cycle,
2. Developing superhuman strength,
3. Loss of appetite,
4. Unnatural body movements,
5. Engaging in self-mutilation or suicide attempts,
6. Sudden divination abilities,
7. Loss of appetite,
8. Becoming a medium,
9. Starting to discuss unbiblical or cult-like topics,
10. Stops blinking their eyes regularly, or their eyes change color,
11. Is a victim of sexual abuse or sexual trauma, pornography addiction?
12. Changes in behavior that don't suit a particular personality,
13. Using drugs or alcohol to cope with pain,
14. Attempted or contemplated suicide,
15. Severely depressed or feeling hopeless,
16. Self-abusive behavior such as anorexia, etc.,
17. Ancestors committed murder, suicide or sexual molestation,
18. Hearing voices tell a person to do immoral or harmful acts,
19. Asking Satan to exchange life for something (fame, sex, etc.)
20. Multiple phobias or fears,
21. Abandoned by parents,
22. Disabling sessions of depression or rejection,
23. Failed repeatedly at significant relationships,
24. Serious health issues,
25. Financial issues,
26. Hindered in Bible reading,
27. Avoiding church prayer services,
28. Gets angry or frustrated if the name Jesus of God is spoken,
29. Loss of hearing or speech functions,
30. Unstable demeanor.

Some demons smell like rotting flesh or sulfur. Some don't. A clicking sound during a paranormal encounter may indicate two paranormal entities communicating with each other.

NO EXCEPTIONS!

I tell a client if an animal or human is receiving violent acts from paranormal entities that they must call the church. I offer no exceptions. I will still investigate and assist.

It's no different that if a human is harmed by another human, someone calls law enforcement.

CURSES

There can't be a quality book about the paranormal book unless there is a mention of demons or curses. This is the other side of the inside of curses.

Implicit curse - The inexorable and relentless, immutable consequence of conduct and communication inherited transgenerationally, continuing until altered by spiritual intervention (incapable of change). Examples are the Irish who have a problem with alcohol, Germans with wars etc.

Explicit curses - The malediction of evil invoked by ritual, ceremony, or incantation with directive intent to bring evil, misfortune, harm or injury to individuals, families, institutions or legacies. It is a prayer for injury or harm.

Curses do not always work just because one person hates another.

Curses are hereditary, involuntary or intergenerational. A curse may never end up as a full-blown possession.

The Bible indicates curses can follow a family or national line for at least 4 generations. But it does indicate that a curse can remain activated after 4 generations, but not active. Exodus 34:6-7.

A curse affixed by a ritual can be removed by the same ritual.

Traducianism- The soul and body come from the biological parent. This theory explains how curses are passed down from generation to generation.

Soul creation - The soul is a direct creation by God. The Spirit is the God-breathed part of each person. The body is the human encasement of all physical actions and sensations. The soul is the individuated aspect of each human being, the sentient, self-perceptual part of who we each are, and that which identifies us as unique, one-of-a-kind creations of God.

There is disagreement between Soul Creation and Traducianism.

Bob Larsen – "Curse Breaking Freedom from the Bondage of Generational Sin"

TAKE YOUR VICTORIES

A year and a half investigation for a family of 8. Mom, Dad, two brothers, four sisters. It included a priest, deacon, bishop, exorcist, a medium, a college professor with some paranormal and psychology experience, an investigation team and lots of neighbors and friends trying to help a family remove a female child ghost with a hatchet lodged in her shoulder blade.

Eventually, over time it was inadvertently discovered that the some of the males in the family were performing incestual sexual acts against the family females. Social services got involved, and child protective services got involved. Law enforcement got involved. The family was split up, divorce paperwork was filed, and the home was abandoned.

I sincerely hope that the family is in a better place now than back then. As far as I know the female child ghost has not attached herself to any of the family members. I can also assume that this female ghost who had a hatchet in her shoulder suffered trauma while alive and remained in this realm and this house because the child ghost wanted to protect the children who resided in the house after the child passed away. This type of ghost would be a purposeful ghost or a protector ghost.

This isn't the type of victory any paranormal investigator wants.

What I am eternally thankful for is that the abuse came to a screeching halt.

I'll have a victory parade for this investigation. It is what it is. The child ghost may still be there. We all know no paranormal investigation is perfect and that with a team that large, there were most likely arguments on what to do or not to do, what to say or not say and when to say it or do it.

The experience taught me never to take anything for granted.

If anything, else, the bravery of those involved is duly noted.

"THE OTHERS"

They may be our competitors or colleagues. The community may have several paranormal investigation teams for clients to choose from.

A woman from a neighboring investigation team asked me to lunch because she saw my investigation report and was impressed. Lunch was great, but then afterward she became bitter and disagreeable with me about my methods and research. She told me she didn't want to work with me in the future because we didn't agree, which was interesting because I don't agree with many people. I investigate a site and provide the client with the investigation results. The investigation results include items that people have different opinions on. The client ultimately decides what to agree with. I'm not the client's mother, attorney, or financial advisor, or God.

Since she didn't want to cooperate, I invited some of her investigation team members to other investigations I did. I pay investigators on my team as soon as they arrive. To my surprise, she was upset that I contacted her team members "...behind her back...."

This woman's paranormal team was her extended family and a primary social support structure for her life. Like a mother protecting her newborn baby, she thought she was protecting her "team", but from what? Just like her, I didn't have all the answers, but I didn't see any reason why members of her team couldn't go on an investigation with another team.

A good supervisor or leader never turns down free training for subordinates. The more investigations people go on the better we get

We need to put the public before our squabbling and cliques. I love this woman just as much as I would love any human being as a fellow human. I hope she does well.

Don't make the same mistake. Share your research, work as a team, tell the client the different views, and let the client decide what is in their best interest.

Human beings are being tormented and abused by paranormal entities. No one has time for pettiness.

Lead, follow, or get out of the way.

THE SOUND OF MUSIC

The evidence is replete. Humanity knows that the paranormal exists. Showing people over and over again that they exist doesn't do society any good if paranormal professionals aren't providing the public with resources to remove violent entities from the lives of victims.

I think about paranormal documentaries. Those made by Hollywood and those made by private paranormal investigators and ghost hunters.

The videos contain creepy music and creepy images that instill fear. Fear sells - I get it. It's a stereotype.

When someone hears the word "ghost hunter" or "paranormal investigator" automatically our minds shift to creepy documentaries with creepy images of ghosts, cemeteries, knives, candles, bats, old buildings and flashlights. The creepy music is in the film.

We aren't going to get communities involved in removing violent paranormal entities if we are scaring them away with our videos and creepy, scary music. Not every paranormal encounter or investigation is a hyped-up Hollywood scream and drama fest.

We aren't going to be able to recruit future generations of ghost hunters and paranormal investigators if they are being scared off by creepy music and creepy videos full of sensationalism, guessing and speculation.

Most videos don't show the investigation team actually helping the client remove an entity. None of the documentaries show the "behind the scenes" work an investigator does such as sitting at their desk for six days straight after the investigation, going over the video footage, audio files, and photographs to see if anything was missed.

The documentaries don't show the weeks an investigator spends alone listening to the investigation that they recorded to see if there were any whispers or voices that were spoken but might not have been heard during the ins and outs of an investigation.

If I was king of the world for a day the paranormal video would have some scientific research, not just an investigation team running through an abandoned hospital proving that there is a paranormal entity there.

Our credibility and ability to get funding from whoever depends on how we sell the paranormal profession to the public in our videos and investigation reports. Our videos sell gore, and we don't provide an investigation report.

How can we, as the paranormal investigation community sell our ideas in an abandoned building occupied by ghosts? A place where no one goes to, no one is getting harmed, and the government can have the building torn down if the building has code violations? The abandoned building can be torn down by real estate developers.

Don't get me wrong. It's a free country and a free society. Publish what you want. What we publish has to add value besides the value of entertainment. The first client who sues the paranormal investigation team for not helping the client remove a violent paranormal entity from the home is going to win. The only thing the paranormal investigation team provided the client was a video with gory music, skulls, cemeteries, creepy photos, and a few seconds of 3 or 4 paranormal encounters. Nothing in the final investigation video helped the client.

The paranormal team made money off the client, the video, and the T-shirts, the promotional items the ads generated and the like. Investigating cemeteries and abandoned buildings is secondary to investigating a home where humans are being violently harmed by a paranormal entity.

HELP IS NOT COMING

Many, many people ask how come governments, churches and scientists and other organizations aren't investing time and money to collect relevant information on the paranormal and UFOs so that we can put a dent in the mystery that shrouds these encounters and entities.

Easy.

Until humans start getting hurt or lives are taken, governments aren't going to invest money in things that disappear and reappear at will. They aren't going to invest in people they think are crazy. They aren't going to invest in theories that beings from other dimensions from other realms are planets are here, and these beings from other realms are doing humans harm.

They shouldn't.

Bigfoot has been around for decades and governments aren't putting any effort into catching the mammal-looking thing. Bigfoot gets away with it because he's not harming anyone. It's that simple.

So, when I hear people say they are upset because governments aren't throwing money at the paranormal problems I'm thankful for things that the government does throw money at. Things like a well-armed and well equipped lethal military force, a police force, a responsive fire department, a helpful ambulance crew, social safety nets for vulnerable citizens in society, roads, bridges, schools, street lights, democratic elections, jails, hurricane and natural disaster relief, dams and other myriad of things that governments provide to better society.

Until these entities start harming or taking lives, we, as a paranormal community, are on our own, and we have to lead, set an example and do our best.

If the governments aren't going to allocate resources to protect the living from violent entities, then it is our job, our responsibility and moral obligation to protect them with the resources we have.

The light at the end of the tunnel is that there are billions of cell phone cameras and video recorders on the planet Earth. Billions of people are recording their paranormal encounters and posting them on hundreds of social media web pages around the globe.

There is no way any government or religion or agency can confiscate billions of cameras and billions of photographs and videos to conceal.

Time is on our side. In 1947 at Roswell, New Mexico, when the UFO landed there. The U.S. Government took the UFO and the photos and the evidence elsewhere. The U.S. Government told the citizens they saw nothing.

The primary paranormal investigator is no longer the government or religious entities. Its private citizens.

The tables have turned. Be patient! Don't give up! Time is in our side!

THE PURPLE EQUIPMENT

When I go to a paranormal site, I determine where the equipment will be set up, where the restrooms are, how many people can fit in the location, where we can park and other logistical details. I may do a pre-investigation site survey.

The primary reason I keep an inventory list of equipment is so I can send it to the client and they can ask me about what I am bringing into their home. If their belief system doesn't want me to bring something into their home I won't. We can also figure out the power and electrical requirements for the gadgets, how many spare batteries I need, how many electrical power strips, and how big the table is and those other details.

If there are multiple teams, everyone will want to know who is bringing what so we don't duplicate where we don't need to, but duplicate where we do.

All of my paranormal gear and gadgets are painted with the color purple. If there are multiple teams on one site gear and equipment can get mixed up, especially in the dark. Everyone knows my gear has purple on it.

If the team leaves gear or gadgets at the clients' house and the client calls me to tell me it got left behind, I know that if the gear isn't purple colored, it's someone else's.

This list helps with my homeowners' insurance claims if my home catches fire.

Equipment Inventory & Electrical Requirements		
Item	Quantity	Power Source
Digital Voice Recorder	1	USB rechargaable battery
Thermometer/Clock	1	
Digital Camera w/ night vision	1	
Digital Camera battery	4	
Pistol Flashlight	1	
Air Sample Analyzer	1	
Magnetic Mount Flashlight	1	
EMF Detector	1	
Thermal scanner	1	
Walkie Talkie	6	
UV & Flashlight (small)	1	
UV Flashlight large	1	
Body camera	1	
Yellow colored gas detector	1	3 AA battery
Spirit Box	1	
Decibel meter	1	
Motion sensor with alarm	3	9V battery
Rook sensor (anti cell/radio)	1	
DAS REM Pod	1	
Laser Grid	1	2 AA battery
Gamma Ray/Radiation detector	1	
Infrared Thermometer	1	
SLS camera & tablet w/battery	1	110V each
Pulse generator	1	Various
Underground metal detector	1	2 each 9v battery
20 port USB charging station	1	110v standard outlet
Lighter (long)	1	
Colored chalk sticks (large)	10	

CHOOSY VISITORS

The client and the public might incorrectly assume that a paranormal investigator can, by default, automatically detect, interact and remove paranormal entities just like an appliance repair service replaces a defective part in a refrigerator and all is done. 90 minutes in and out. That's not always the case.

Most investigation teams don't have a medium, psychic, ritual, or the like that can do this right away. Removals may occur on a second visit or be done by someone else.

When an investigation team arrives at a location, the team doesn't know if the entity even wants to interact. The entity may have crossed into the light or left the area. Some entities may be stubborn.

The entity might not want to interact with males but will interact with females. They might or might not want to interact with a pet. A female ghost who suffered trauma from a male might not want to interact with a male but will interact with a female. Some entities will interact with one person or with many.

I investigated a haunted house with a super team of individuals who have been in the profession longer than I have. It was a great experience and we had unrestricted access to the manor and plenty of snacks!! The last time the investigation team was there, about 4 years ago, I wasn't there, but the team had excellent interactions with the paranormal on their last visit. This visit, however, produced fewer interactions. Everyone was disappointed. The team members were family members who traveled long distances to investigate. There is no glory or fame in coming up empty. We had fun, talked, ate, shared stories, tested our equipment, trained new people on the profession, exchanged theories and it was a great experience breaking bread with other people.

Paranormal entities typically don't speak and appear at the same time because it requires so much energy to do so. Their interactions are brief so an entity might do one thing or another, but rarely two things at the same time. Despite the stereotype that everyone thinks they only speak they may also:

- Knock
- Scratch
- Move something
- Slam a door
- Open a cupboard
- Tap (including tapping a human)
- Touch a human
- Walk across a squeaky floor
- Move a curtain
- Fiddle with an electronic device or battery-operated gadget
- Touch hair
- Touching the neck or back

We automatically assume the paranormal can fluently speak, read and write in our native language.

WHO IS THE TALKER?

There is nothing more exciting than to get a response from an entity when the investigation team interacts with it. A common mistake investigation teams make when trying to verbally communicate with a paranormal entity is that:

1. The team may not know how many entities are in the site.
2. Several team members ask a question at one time and multiple questions may lead to multiple answers and therefore confusion.
3. The entity may not have enough energy to respond to the question and may provide a short answer or a clue for an answer that the team is to decipher on their own.

If only one person at a time speaks, then things go smoothly. A male can speak, then a female, an older investigator or younger investigator and maybe the client.

Just like living, successful communication begins when one person speaks at a time uninterrupted.

The ghost might wait for permission to speak.

I'm not trying to be stereotypical. Some ghosts don't want to share their feelings, just like living humans.

Instances had occurred when a ghost was asked about sexual things that weren't always pleasant in everyday conversations, but they answered.

Thank the ghost for interacting. Just like Mom said, use your manners and say please.

BINARY QUESTIONS

Our paranormal friends, if they desire to interact with us verbally, have limited amounts of energy to speak, and we don't always know the mental condition, purpose or agenda of the entity we verbally interact with.

Therefore, we must ask questions to the entity to elicit a specific response but a response that uses the least amount of energy for the entity to expend to answer us. This is achieved by asking the entity either a "yes" or "no" question or a "binary question." A binary question is a single question that provides more than one answer. The answer may even tell the investigation team the purpose of the entity and why it's here.

It is most positively critical that we tell the entity ***HOW*** we want it to answer such as "blow out a candle, move something, knock, scratch, touch etc.

Here is a list of "binary questions."

1. Are you female?
2. Are you a follower of Satan?
3. Were you ordered or forced to stay in this realm?
4. Are you able to cross over to the light?
5. Do you want to cross over to the light?
6. Can I help you cross into the light?
7. Can I do something to give you peace so that you can cross into the light?
8. Were you sick before you passed away?
9. Did you commit suicide?
10. Were you murdered?
11. Was there an accident?
12. Do you fear punishment by God if you cross over into the light?
13. Do you fear arguments with others already in the light?
14. Are you here just to prove to the world that you exist?
15. Do you have a message for someone?
16. Were you summoned here by someone else?
17. Are you here because of a curse or hex?
18. Can you travel to this realm at will?
19. Are you here for a specific person?
20. Are you here because you like a specific object or thing?
21. Are you here because you are being punished?
22. Are you here for asylum?
23. Are you here to protect someone?
24. Are you here to apologize?

WHO DO YOU TRUST?

One of the most common questions I get is if the paranormal entity is something we can trust.

Trust the entity until it gives you a reason not to. Depart from it immediately if it starts out as friendly then turns malevolent or interacting with it becomes unsafe.

OUIJA BOARDS & SEANCES

The Ouija board is another symbol of the stereotype of the paranormal profession.

The mistakes:

- More than one entity appears but may not interact.
- All may not leave when you are finished.
- The users of the board forget to ask the entities to leave and not attach.
- Many users forget to state the session is over.
- Someone may have used the board before the current occupants of the home occupied it, and things were already there.

I respect every human being that is a medium or psychic. People have gifts, and not everyone uses their gifts for good. Anyone can do what they want with their free time and believe what they want to believe.

No disrespect to those who have managed to make the Ouija board work for them, and they haven't suffered long-term negative consequences for using it.

I had a client who was using a Ouija board and made all the mistakes. He called me because he woke up with bloody ankles, bloody sheets, and bloody steak knives at the foot of his bed. I can't imagine what he told his wife or children or what the dog thought, or what the carpet cleaning bill was.

I can't make this stuff up. There are thousands of stories of things going wrong with this board. Those stories speak for themselves. I've even had people tell me they used Ouija boards with their parents and grandparents 30 years ago and they are still having problems.

The Ouija board speaks for itself.

I ask people to hold one more Ouija Board session or séance and then throw the board out, rendering it useless.

If we don't want harmful things in our home, we don't purposely invite them into our home.

WHEN IS IT OK TO BELIEVE?

I'm not bothered that people don't believe in the paranormal.

People who experience it first-hand didn't believe it either until they were forced into it.

Some didn't believe in UFOs until they saw one while driving home from work late at night.

Bigfoot too.

I won't convince them to believe. That's not my place.

The universe will unfold as it should. It always has and always will.

Everything happens for a reason, good or bad.

I will try to help those who ask for help and aren't unreasonable with their request.

The best way to be the bravest person possible is to undertake something that terrifies you.

TESTING 123?

I read a book, and I won't say the author. He suggested that those who invest in paranormal electronic gadgets not use the gadget in their home out of fear that they may find a paranormal entity.

I totally disagree.

Chances are the client who contacted the paranormal investigator is at their wits end and they have run out of options and they finally break down and call an investigator. They don't want to invest hundreds of hours researching the paranormal online or buying books or trying rituals to remove what they don't know is there. They don't want to invest thousands of dollars in electronic paranormal investigation equipment.

The customer/client expects and deserves a brave paranormal crew in their home.

I get it if the paranormal investigation goes horrible and everyone gets scared or someone gets injured and people want to leave the investigation site for the personal safety of the team, then by all means depart quickly and safely.

I agree that a gadget owner can take their gadget to a cemetery or outdoors or someplace else and test it and configure it. They can also test it or configure it in front of a client. An investigator looks foolish in front of a client if the gadget doesn't work or the investigator doesn't know how to use it.

Bravery is a mandatory skill and requirement in this profession.

If a gadget owner is too scared to use their own gadget in their own home and that they spent their money on in the privacy of their own home then that just means that they are too scared to use their gadgets for a client.

No one buys a thousand dollar anything and then has no expectation of using it unless it's a gift.

The paranormal gadgets are gifts to you and gifts to your client. The gadget is your helper.

Every gadget has a purpose and every gadget has its pros and cons.

Home is where we practice. Home is where we calibrate and test before we deploy the gadget to the rest of the world.

Please do the paranormal community as a whole and stop being a paranormal investigator and donate your gadgets or sell them on the Internet if you're not brave enough to use them in the privacy of your own home.

No one in the paranormal profession on the planet Earth needs a paranormal investigator who lacks bravery.

For those who lack bravery, I`m not negatively judging you. Every person has their skills, abilities, talents and faults and quirks.

The unbrave can still help. The unbrave can do office work, schedule the team, arrange for transportation, order shirts and supplies for the paranormal team, cook meals or prepare snacks, edit and upload videos and photos, answer the phone, set up and maintain the company web site. Do research and find other ways to support the client and the paranormal team.

But for God's sake a paranormal investigator has to be brave. Just like in the military at war, a weak, scared soldier ruins it for every other brave soldier, and no group of brave soldiers wants a single soldier that is scared in their military unit. It is a risk and hazard to have an unbrave person on the paranormal team. I've asked people to leave who freaked out too much. They're human and chances are it was their first investigation.

If the "first timer" tried it, got too scared and didn't like it then that's OK. It's not their calling any more than it's a plumber's calling to deliver a newborn baby. Some people just aren't wired for it.

If you're an investigation organization, don't judge. Let others contribute to the effort using their skills, talents, and abilities. Everyone can contribute something.

USELESSLY USEFUL GADGETS

Never before has it been easier to track and learn about the paranormal

- Electronic gadgets can detect them.
- A wide variety of internet research sources are available
- The ability to research or be at an investigation site via social media means we learn in the comfort of our home and never leave home, even if the investigation is in another country.
- Our gadgets extend and amplify the five senses of humans (sight, scent, smell, hearing, touch).
- The gadgets are low cost enough to perform functions that humans and animals can't.

A drawback to our gadgets is that a 300-year-old paranormal entity might not be able to relate to a 21 century electronic gadget or interact with a 21 century gadget. It's not familiar to them. They may think they are going to get trapped or harmed if they interact with it.

The investigation team doesn't know the competence level or I.Q. of the paranormal entity.

The investigator may have to resort to old-fashioned items to "trigger" the paranormal entity to interact. Instead of using a motion sensor or laser grid an investigator will also want to add a baseball, a doll, a candle, and maybe a toy car to their investigation inventory and deployment.

I conducted an investigation that was going quietly until someone added a cigarette to the scenario. Then we got paranormal readings off the charts after the cigarette was presented was announced. We are assuming that the ghost missed smoking cigarettes and once it saw the cigarette we were "smoking friends" and then the ghost interacted freely.

If we want to communicate or interact with a 200-year-old paranormal entity, we might have to use gadgets, toys, and household objects, and things that were around in their time period-200 years ago.

I will listen to any investigator that doesn't like a particular paranormal gadget. The question I ask them is what gadget can they bring to the table that works better?

I will sit down with a fellow investigator and help him or her write a letter to a gadget manufacturer on how to improve a paranormal detecting gadget. When we both write the letter, we aren't arguing with a peer and we are both working together to enhance the profession.

There are no perfect paranormal investigation devices. Each device has its pros and cons, and each may serve a different purpose.

Send feedback to the gadget manufacturer. The manufacturer wants to produce and sell something in the field that is useful and they appreciate it when those who are on the front lines provide feedback. The manufacturer won't make money unless the gadget provides the investigator with what the investigator needs and wants. It's free enterprise, supply, and demand. It's a perfect system.

WHY WE NEED A PARANORMAL BODY OF KNOWLEDGE (PBOK)

Modern technology and media have reached a point where we know for a fact that paranormal entities aren't a figment of imagination anymore. Any reckless citizen or group in society such as those who take advantage of others don't add value to society, steal, damage things, and those refuse to help others but have the ability to act and don't are the ones that lose. As a collective group, the paranormal community must be organized to ensure safety, get public support, solve difficult problems and work together. The only way our profession is going to thrive and survive is if we as a profession know what we agree on, what we disagree on and focus on what we agree on. Just having a paranormal investigation team show up and say "Yep, you have an entity" and then walk away from the client looks poor for our profession, especially if the client is being harmed by an entity. Just like medicine, or law, if we collect our data and research and it's coherent, realistic, organized and solves problems in society, more people will accept the profession, participate in paranormal interventions and contribute positively to society..

What does the PMBOK look like? Here is an image of its initial structure:

The Paranormal Body of Knowledge (PBOK)				
The Entity	**The Humans**	**The Environment**	**The Technology**	**The Investigation**
Active	Victim	Colors	Gadgets	Methods
Residual	Investigators	Objects in home	Holy Water	Pre-Investigation
	Family/Friends/Neighbors	Location	Oils	Actual Investigation
	Clergy	Antiques	Ceremonial Objects	Post Investigation
	Mediums/Psychics/The Gifted	Dolls	Ritual Objects	
	Scientists/Mental Health	Ouija Boards	Media	
	Cults and Organizations	Electronics	Equipment/Machinery	
	Séances/Ouija Boards	Animals/Pets		
	Séance Members	Urns		
	Curse Givers/Receivers	Ceremonial Objects		
	Laws and Regulations	Ritual Objects		
	Religious/Spiritual beliefs			
	The Media			

The medical profession studies and documents of the human body by body parts and the internal and external organs of the body. The legal community studies and documents the legal profession through public law, civil law, criminal law, constitutional law, family law, and the like. So, must the paranormal community at large.

I'm not saying this chart is perfect or all-encompassing. There may be things to add, duplicate, and remove or move. Just like finding cures for diseases, or finding the appropriate medicine to treat an illness the medical community collects data and has information and processes and systems in place, and still does to perfect medicine. We have to get past the "we don't know" and the sensationalism, the gore, the scary videos with creepy music, the skulls, the cemeteries, the knives, and the rest of Hollywood that have consumed the profession that our investigations are mere entertainment and not helpful.

I've seen first-hand a paranormal client who is being tormented by a paranormal entity. They despise paranormal documentaries and YouTube videos about the paranormal because the documentaries don't show paranormal investigators or ghost hunters removing violent entities.

We want to study something with others who believe what we believe in. Not argue with those who disagree with us. When a person goes to college and takes a class, everyone in class is studying X and, in the classroom, next door everyone is studying Y.

There is no better way to figure out ghosts than to share the experience from those who have experienced it firsthand. The perspective of how a ghost interacts is better from the perspective of a human being who has seen ghosts all of her life, could speak to them, and even remove them on command.

THE FAMOUS "ORB"

When I peruse social media. I see hundreds of photos and videos people post of orbs. Orbs are tiny balls of light, sometimes as small as a pencil eraser. Some are different colors. Thankfully, I've never observed or heard of anyone being harmed by an orb. I see them all the time on investigations.

The number one disagreement on social media is if the orb is actually an orb or if it's just a reflection or some other light source of if it's glare from the camera. This is where we go off the deep end as a paranormal profession. The arguments last for days and worse yet, not a single person on social media has actually gone to the site to investigate where the orb was spotted.

We know orbs exist. The evidence is clear. But we keep asking the question if the photo or video is an orb or not.

We have to move past that and the arguments and disagreements and egos that surround it. Instead of asking if the photo or video is an actual orb or something else let's ask:

-Is the orb an energy source for another paranormal entity?
-Does the orb travel by itself or one or more other entities?
-Does the orb provide an energy source for other entities?
-Is the orb protecting us?
-Is the orb a human astral traveler?
-Is the orb an angel or violent?
-Does a paranormal entity need an orb to exist?
-Do orbs interact?
-Do they travel between realms?
-Are orbs, leaders, followers, or bystanders?
-Do orbs have a scent?
-Do they require or give off energy?
-Does the orb have an inner and outer shell?
-Is there something else living in the orb?
-Does the orb manifest to something or does something manifest into an orb?
-What is the temperature of the orb or is temperature influenced by the orb?
-Is the orb conscious like a human or conscious like a rabbit or worm?
-Is the orb influenced by electricity, chemicals, heat, light or other environmental conditions?
-Is the orb man made or naturally made?
-Can an orb be created or destroyed by humans?
-If we destroy or remove an orb will is destroy another entity?
-Does an orb appear before, during or after something?
-Are orbs tied to humans or animals or "attached"?
-What is it made of?
-Solid? Liquid? Gas?

You understand?

THE ZOO AND FIDO

To perfect the paranormal investigation and our research it's challenging in that humans can capture wild animals and transport them to the zoo, lock them up and study them to our hearts content for decades. Not so with the paranormal. Paranormal entities appear and disappear almost at will and unless they are a residual ghost that doesn't know it's appearing and disappearing the investigator only has a few seconds to capture and retain as much data as possible before the entity disappears.

Speaking of animals. Our nearsighted dogs and cats and other pets don't have an ego. They operate based on instinct and emotion. Which is why they don't question or second guess if they sense something paranormal that a human can't detect. Fido the dog has good hearing and he may hear a tiny piece of plaster or sheetrock fall from behind the wall and think it's paranormal, and you have no idea what Fido is barking at. Dogs have a natural curiosity in new places as well.

Their whiskers, sense of smell, sense of hearing and eyesight are fantastic paranormal tools. If you can get permission from your client, bring the dog as an organic paranormal detector. Labrador retrievers have the best eyesight if you compare them to all other dogs and the 20/20 eyesight of humans.

Our dogs have a tapetum lucidum. It's located behind the retina, the tapetum lucidum reflects light through the retina, increasing the light available to the photoreceptors which is why dogs see better at night.

Dogs have more rods in the retina than humans do. Rods are sensitive to shape, movement, and dim light. Dogs can see moving objects much better than stationary objects, and they have 10-20 times greater motion sensitivity than humans. They have more rods than cones in their retina. Larger pupils that allow more light into the eye and a lens that is closer to the retina, making the image brighter.

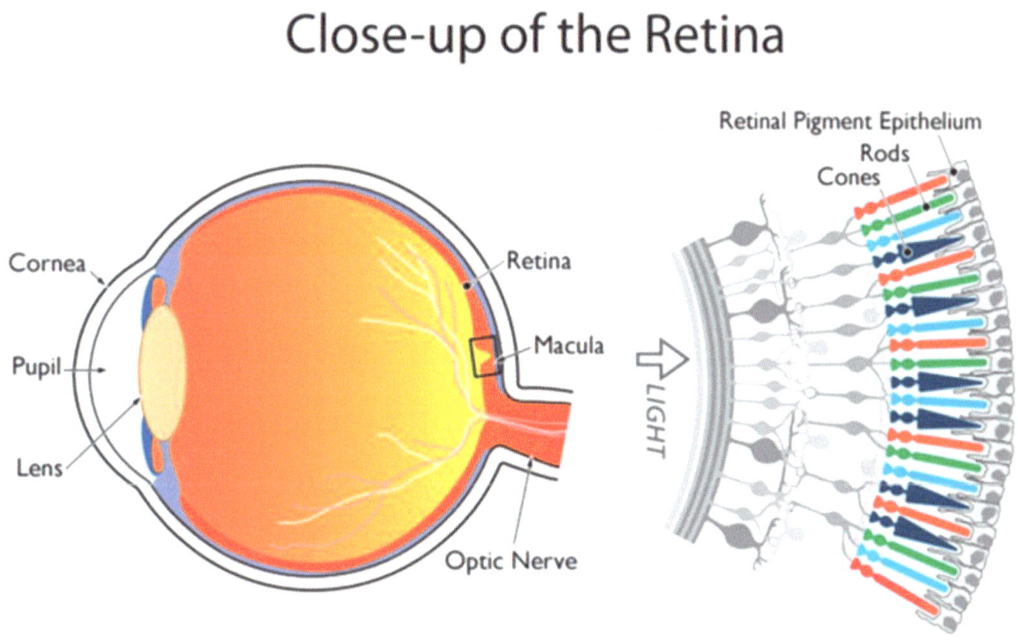

Source: https://www.winetasting-demystified.com/sense-of-sight.html

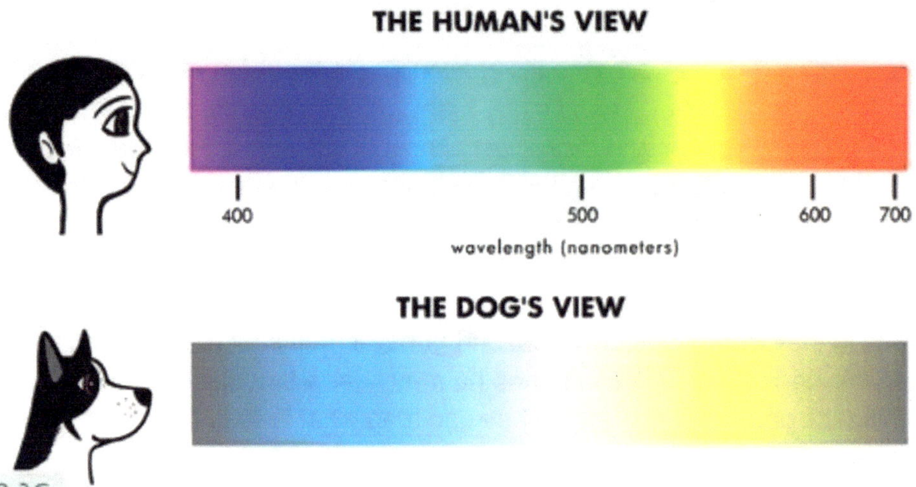

Fig. 1.2.3C
Dogs have two types of color photoreceptors while humans have three. As a result, dogs probably perceive the world in different saturations of yellow or bluish-gray.

https://www.quora.com/How-good-are-dogs-vision-compared-to-average-humans

PART 2

WHERE THE SCIENCE MEETS THE GHOST

I can't stress enough that the purpose of the paranormal investigator is to **prevent or stop violent entities from harming the living.**

I can't stress enough that if ghost hunters and others are going to get sloppy and not respect laws and shortchange clients, then the legislatures are going to put laws into place on the paranormal profession.

The church and science are not going to help us, and if they do, then they are very limited in their ability to help or want to help. We're on our own.

Therefore, at this period in human history we must determine how a ghost manifests itself so that we can prevent it from harming humans.

Humans are being told to wear crucifixes and garlic, carry Bibles, or sage all with mixed results.

How does the entity manifest? How can we prevent its manifestation so it won't harm the living?

That is our focus. That is our calling and mission.

What we know is that ghosts always appear in the color white and only the color white. There must be a reason why this color is chosen or used.

WHAT OTHER BIOLOGICAL BEINGS ON EARTH OR WHAT OTHER NATURAL PROCESSES OCCUR ON EARTH THAT APPEAR, GLOW OR PRODUCE THEIR OWN LIGHT? SPECIFICALLY, THE COLOR WHITE?

A lightning bug or a firefly? A flashlight? A lightbulb? A flare? A flame? A star? A jellyfish? A meteor? A glow stick? An X-ray machine? Clouds? the mist that comes out of the mouth of a human when the human exhales in cold weather? White caps on moving waves or water? Dyes and colors? Sparks? Fog? Bubbling hydrogen peroxide? Salt? Sugar? Fog from a fog machine? Car battery corrosion? Strobe light? And all other things that naturally appear or are made to the color white.

My research assumes that Mary Ann Winkowsi is correct in her book "When Ghosts Speak: Understanding the World of Earthbound Spirits" ISBN: 0446581186, ISBN 13: 9780446581189 published by Grand Central Publishing. Her life's work was also the driving force behind the television series "Ghost Whisperer" by Jennifer Love Hewitt. Her book and the TV show is a classic reason why we can and should share our paranormal experiences in a PBOK to solve the problems our paranormal friends throw at us. First-hand experience is golden. I'm sure the TV show was a bit hyped up, as most are, but the show is as insightful as the book.

She said she was able to "make the light" that ghosts could walk through to leave this realm, and therefore stop disturbing the lives of those who had unpleasant paranormal encounters. She has spent her life working and interacting with ghosts and earthbound spirits. She believed that when humans die, we go to our own funeral, then are able to cross into the "light" within 72 hours. Young children who didn't have the competence to cross over were usually aided by a grandparent or angel to cross into the light. Some who don't cross into the light have specific reasons we already discussed.

She was spot on when she said that if a human was friendly while living, the human is a friendly ghost. Conversely if the human was horrible, their ghost version is horrible. She is correct that the ghost acts their age and biological gender.

In my own research a child ghost appeared in the bedrooms and playrooms where living children played, including outside playgrounds, and backyards. Adult ghosts appeared in kitchens and garages where adults are, more than children.

She also stated that the living can talk to the dead, but the dead must cross into the "light" and the dead can only speak to the living, when the living human is asleep. She believed that it's very difficult for a ghost to interact with the living unless it crosses into the light, which is why we have paranormal activity.

Mary Ann was able to convince a ghost to cross into the light, but none of her books described how they manifested or obtained energy to interact, move objects or if there was an energy draw before a paranormal interaction occurred.

The purpose of her book wasn't intentionally written to answer the question about HOW the entity interacted with the living in this realm when the ghost is supposed to be dead.

I'm not blaming Mary for not knowing or accusing her for not sharing if she knows. Because Mary Ann can interact and remove ghosts at will, they are simply handled by her as an annoying guest or children. She doesn't take her ability for granted and clearly by her words and actions she is using her gift to benefit humanity. Mary Ann didn't ask the ghost why a room gets cold, why people got chills or why floors creaked or the why lights flicker or why candles go out as part of her conversations.

I am grateful that Mary Ann contributed to the understanding of the social side of a ghost. She removed peaceful and not so peaceful entities.

Mary Ann invests her time in developing her gifts to their full social, moral and economic potential and value and tells everyone they can "make their own light" to remove a peaceful or violent ghost. Few if any have been able to perfect the ability to make their own light.

We don't have billions of Mary Ann's who can interact with a ghost and remove the ghost immediately. She is mortal and isn't training battalions of people to follow in her footsteps after she passes. She is writing books though.

If Mary Ann and a psychic and medium say they can remove a ghost just because they have a gift and they can, that doesn't always help the rest of us who can't. It's similar to the old belief that a woman carries a baby and makes her own baby. They could make, carry and deliver their own baby "because they can". It wasn't true until it was researched that a man was required. "Because they can" changed to "She had help" from a male sperm and a physical sexual act.

This is what I propose. The Church says the removal of a demon is a divine act. Mary Ann and the mediums say the ability to remove is a "gift". There must be another way.

We all know Mary Ann is only one person and she has a limited amount of time to travel to every haunted location to remove a ghost. She's mortal just like the rest of us. Her methodology is a gift. Her books show us that the way to remove a violent entity is by a person's natural gift, but it doesn't help the reader with ways to remove a violent entity unless they have a gift. We know that violent entities can be removed by people who don't have a gift. That's what an exorcism is. The priest doesn't remove a demon in an exorcism, the priest is the valuable, sacred middleman. The priest usually isn't a medium or psychic, in fact a priest is prohibited from being a medium or psychic or believe in such things as a condition of remaining a priest. A priest who uses a Ouija board, and any other religious device not sanctioned by the Church would be ousted from the Church without question.

I'm looking for a third way to remove a violent entity.

1. The person has a gift (medium/psychic) who may have the stereotypical perception as being ineffective or a snake oil selling kook.
2. A priest during an exorcism, who is the middleman working through God, Jesus, and the Holy Spirit

And

3. Determining how white colored "things" on earth manifest themselves, and how to inhibit or stop their manifestation to prevent violent entities from harming the living.

She said that every person can "make their own light" to remove a ghost and get it to cross into the light but I don't know anyone that can mirror what she does. I'm working with a friend who says she can see ghosts, but hasn't perfected "making the light" so the ghost can "cross over".

Although the ghost may be roaming freely and harmlessly, the paranormal activity living humans experience is when an earthbound ghost wants to interact with the living but has no energy so they interact with manmade or natural energy sources to interact.

I am looking for the biological and scientific structure and order of the ghost in an effort to establish methods, technologies, and barriers to prevent violent entities from harming the living.
I can only ask that the paranormal investigators in this generation and future generations change course. Our grandparents gave us Project Blue Book and they gave us old fashioned cameras. Our great, great, great grandparents gave us their energy and afterlife to present themselves to us as ghosts. Even to this day when our ancestors visit us in the form of ghosts, they help us all become better investigators.

We owe their generation and our future generations to take their baseline and move forward as a body of professionals.

Governments are not going to reveal secrets. The Catholic Church is not going to reveal their secrets. Until they do, it's up to the "mom and pop" investigators to do the research, find experts who are willing to donate their talents, gifts, abilities and education to the same, simple, civil and moral goal of stopping or reducing the occurrences of paranormal entities harming the living.

Again, our purpose for the 21 century and beyond is to collect our data during the paranormal encounter and instead of one or two people giving up their time to look at photos and videos and then making a video to post, we sit down and look at the science behind the ghost. This also means that the paranormal investigation team gives the client something tangible, such as a formal investigation report that they can hold on to in order to make informed decisions. A formal investigation report can also be left behind for future occupants of a house, if the current occupants decide to move out because of paranormal activity and they want the next occupants to know that they are not alone.

If we look beyond white on the planet Earth we know that angels in the Bible were white and of all the chapters in the Bible white is mentioned more than any other color (white 7, gold or yellow 4, black 4, blue 3 and red 2). The grade school pyramid with the rainbow ray of light passing through a triangular prism and making the single color of white is a well-known experiment.

PROPOSED Law of Paranormal Physics 1 (LOPP1)

Any paranormal entity interacting with the living on Earth is subject to the laws of physics on Earth.

Not all laws, but the ones that allow them to manifest enough to influence the 5+ senses of the human and pets, move objects and create a paranormal encounter.

I propose that ghosts manifest themselves by way of thermal reaction, chemical reaction, nuclear reaction, biological reaction or electrical reaction. They may manifest with all, some or none of these processes.

to prevent violent acts of paranormal entities against the living?

Lightning bugs glow, x-rays from the hospital appear in the color white, candles give off light, fire gives off light, a glow stick gives off light, electricity gives off light.

To determine how a ghost manifests, then we can look at other things in nature and how they manifest and create their light.

PROPOSED Law of Paranormal Physics 2 (LOPP2)

Paranormal entities use DNA as a reference point to manifest to humans or animals in earth environments, but they do not contain large amounts of adenosine triphosphate (ATP).

We know that ATP is the growth “thingy” in humans that causes cell development and aging. When a human encounters a paranormal entity (with the exception of a demon or orb) the entity doesn’t appear to age.

This doesn’t limit or exclude the possibility that they may have ATP to manifest, but the fact that a child, pet, or a war hero dressed in military garments from the 1800s and women wearing long dresses that are no longer worn in modern history indicate that don’t age.

If they had a cognitive ability to recognize their clothing, or they had the ability to change their clothing they might. Their clothing could be a reference point for the living.

The beauty of the mystery of clothing on ghosts, and the fact that we can have child ghosts, and animal ghosts is that it provides us a clue as to the age of the ghost, and a time period and maybe a geographical region of where the ghost formerly lived.

PROPOSED Law of Paranormal Physics 3 (LOPP3)

Not all ghosts or paranormal entities have advanced degrees, training, education or competence to understand the basic laws of science or physics. Therefore, paranormal entities have figured out instinctively, or from others in or out of this realm how to extract energy from the earth to fulfill their purpose, or they are given the ability to manifest. We never see a ghost mixing chemicals and compounds with test tubes and Bunsen burners and then “poof!” they appear for longer than a few seconds. They grab whatever energy they can get, from wherever they can get it and use it. Whatever energy they are grabbing they know it works, but it doesn’t work for long and it would seem like there isn’t enough of it to stick around.

PROPOSED Law of Paranormal Physics 4 (LOPP4)
"The Firefly Baseline"

The baseline as to how a ghost appears to a human so that the ghost is visible and the human can hear the ghost occurs when:

- The audio frequency is approximately 175-10550 Hertz,
- The brightness is about .325 lumens,
- The volume level is about 30 decibels or higher
- The color wavelengths in the gamma or cosmic frequencies
- Human brain Beta wave - 12-40 Hertz - Conscious reasoning
- Human brain gamma wave - 40-70 Hertz - Self-aware, happy, learning state
- The interaction is an EXOTHERMIC reaction

This baseline is called the "firefly baseline".

THE SOUND OF MUSIC PART 2

Speaking of sounds. If we can hear our paranormal visitors with or without devices then we can assume that the entity is able to speak at a decibel rating of at least 30 decibels or higher. That is the level when humans can start hearing whispers or tree leaves rustling.

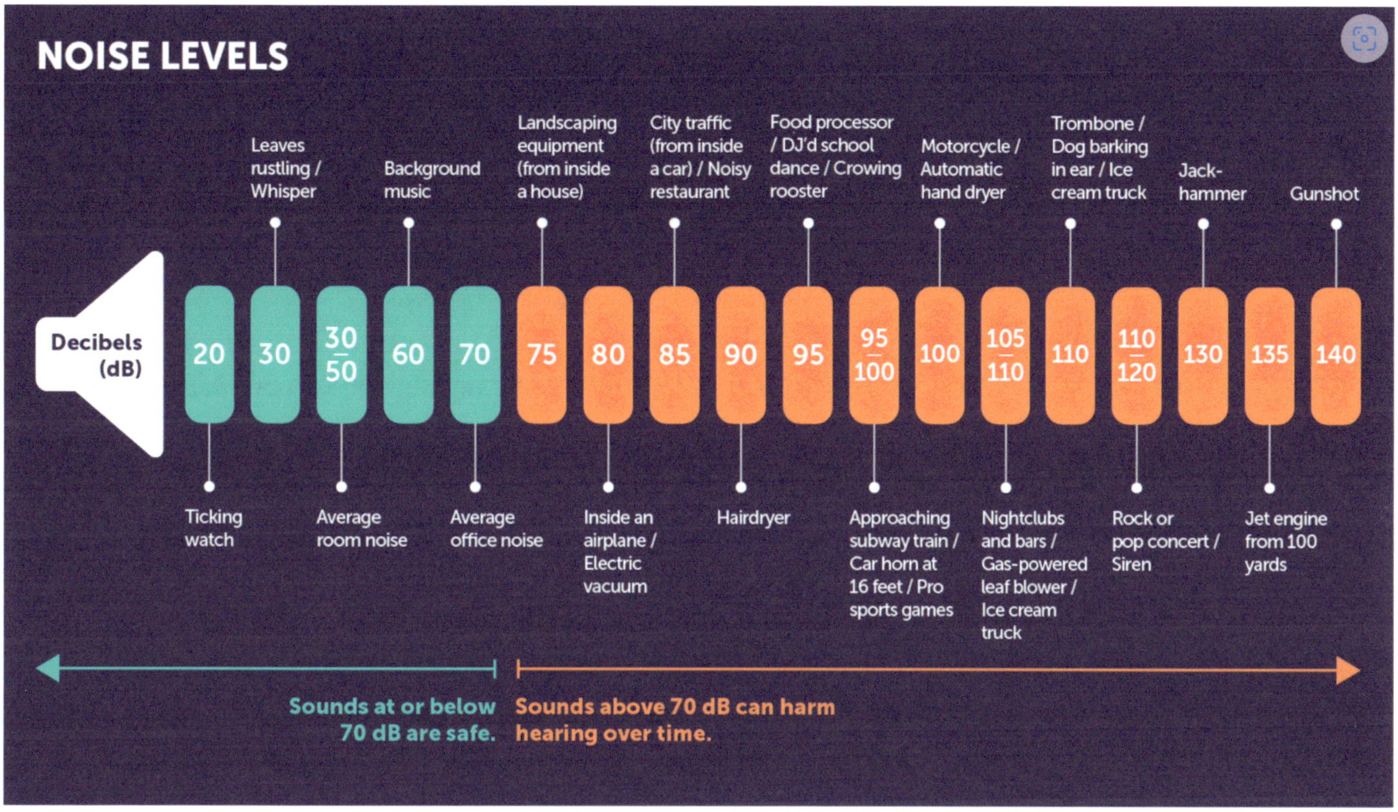

https://hearinghealthfoundation.org/decibel-levels

Our beloved pets hear at higher FREQUENCIES (not DECIBLES).

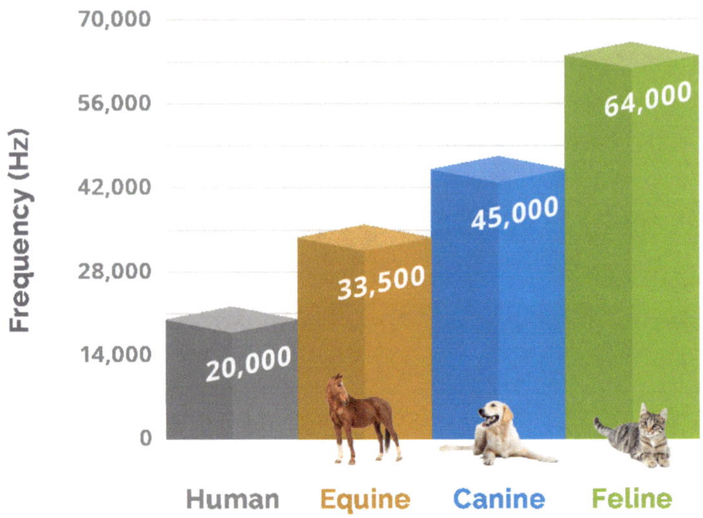

https://thepetcalmer.com/wp-content/uploads/2018/07/1487978927735.png?x85866&x85866&x85866&x85866

Decibels is a measurement of volume, Frequency is a measure of wavelength. There is no device that can convert decibels to frequency or vice versa.

The frequency of a sound is the number of cycles of a sound wave in one second.
The term dB (decibel) and the dB scale are used world-wide for the measurement of sound levels.

DNA, RNA, ATP

The ghost is on planet Earth, so unless it's a miracle, the ghost or paranormal entity must interact and follow the same laws of physics that all organisms follow. Demons smell like sulfur or dead flesh, so there is a chemical reaction to make the demon smell that way, and the human nose has pieces and parts to smell the demon!

The baseline is that the ghost has DNA or RNA as a reference point and that there is some kind of reaction, process or disorder that makes it appear to the living.

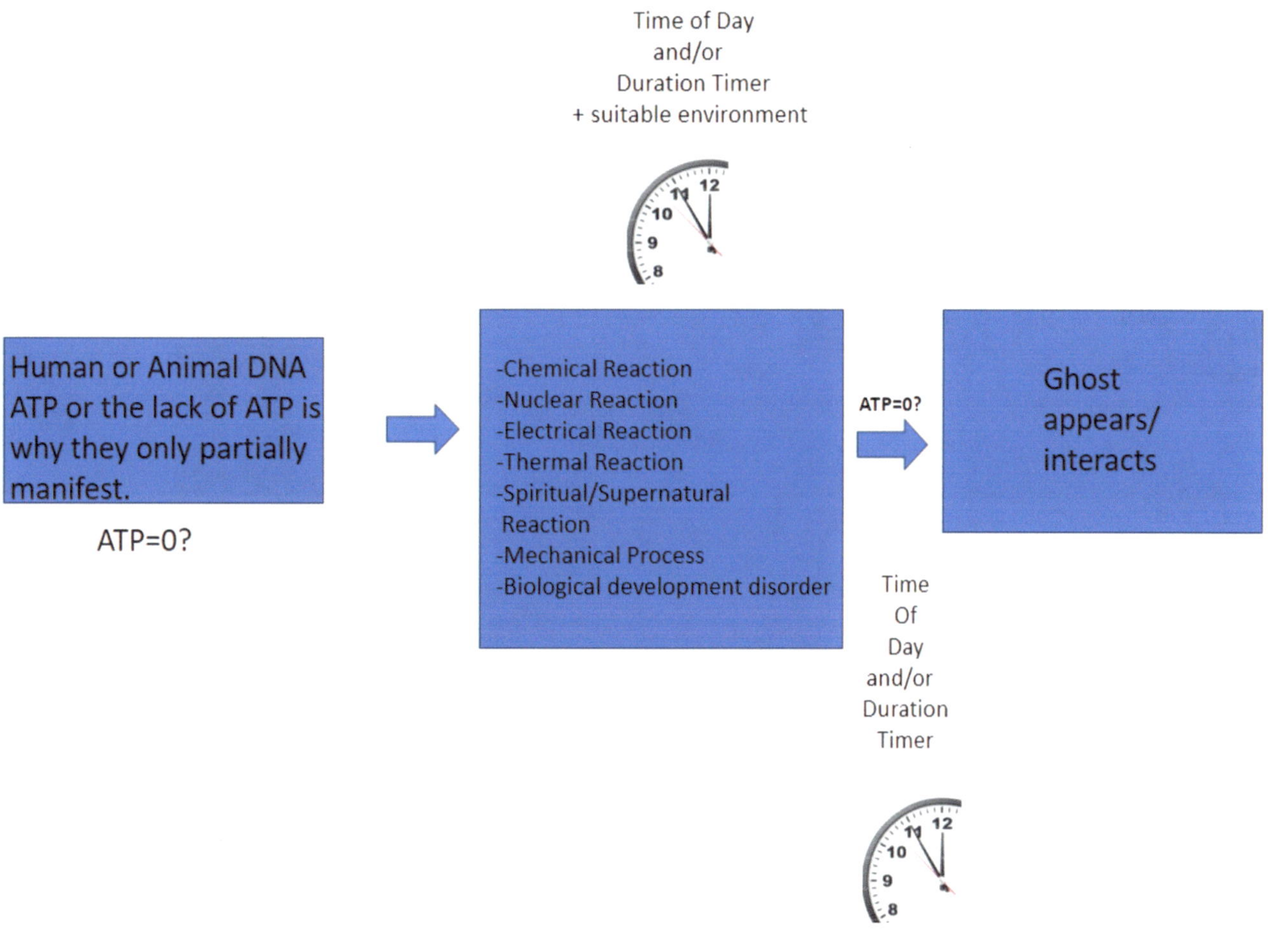

DAUGHTER CELLS

Another theory is that ghosts are entities that contain daughter cells.

Daughter cells are cells that inherit genetic material. So, when the ghost appears it is taking the form of a human or animal that previously had living DNA. The ghost then disappears. When the ghost reappears, it reappears it's still the same ghost regardless if it's a residual or normal haunting. We can also assume that ghosts are daughter cells because they inherit genetic material from the deceased person and may appear more than once (either as a residual ghost or not) at the same age.

This chart represents one possible way, but the daughter cell and the energy taken and used when the ghost appears may move the blue and orange squares in multiples places and positions:

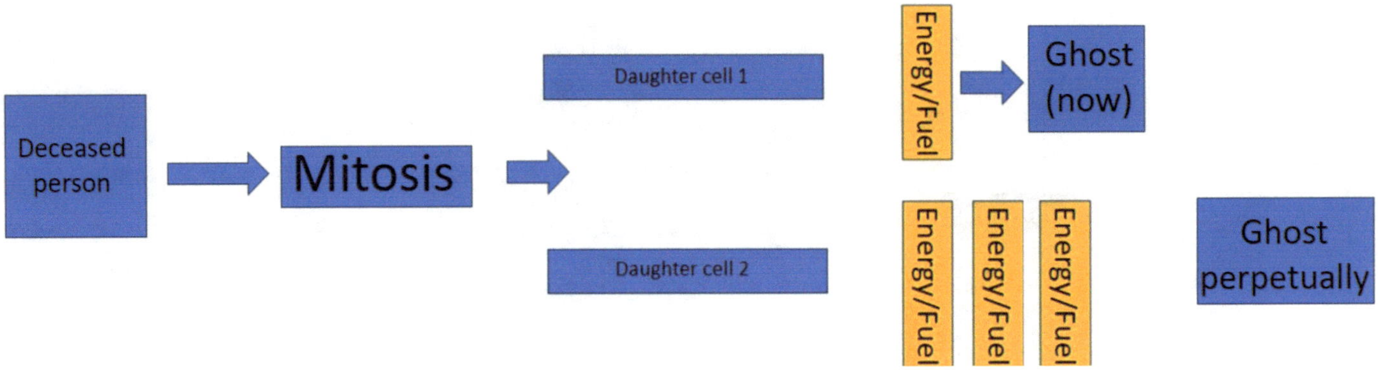

COLOR MIXING

We know that white can be created by mixing colors. White may simply be a camouflage, or the energy required for a ghost to manifest results in the color white:

Additive color

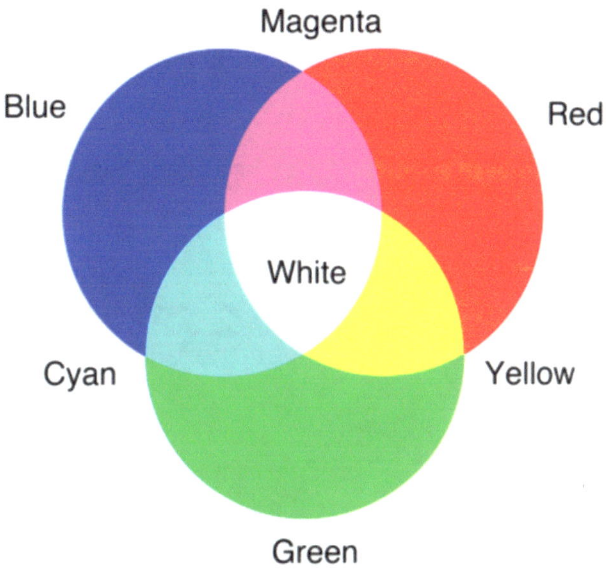

https://marketingaccesspass.com/what-colors-make-white-how-to-make-white/

BRAIN WAVE FREQUENCIES

Brain waves of humans are of a certain frequency depending on the consciousness level of the human:

Delta wave - Deep sleep 1-4 Hertz

Theta wave - 4-7 Hertz Meditation/sleep

Alpha wave 7-12 Hertz - not coherent and relaxing

Beta wave - 12-40 Hertz - Conscious reasoning

Gamma wave - 40-70 Hertz - Self-aware, happy, learning state

We can assume that a ghost that is able to interact with the living uses energy that its frequency is a Beta or Gamma Wave.

The Gamma Wave of an X-ray machine is at a much higher frequency though, between 10^{16} and 10^{19} waves per second.

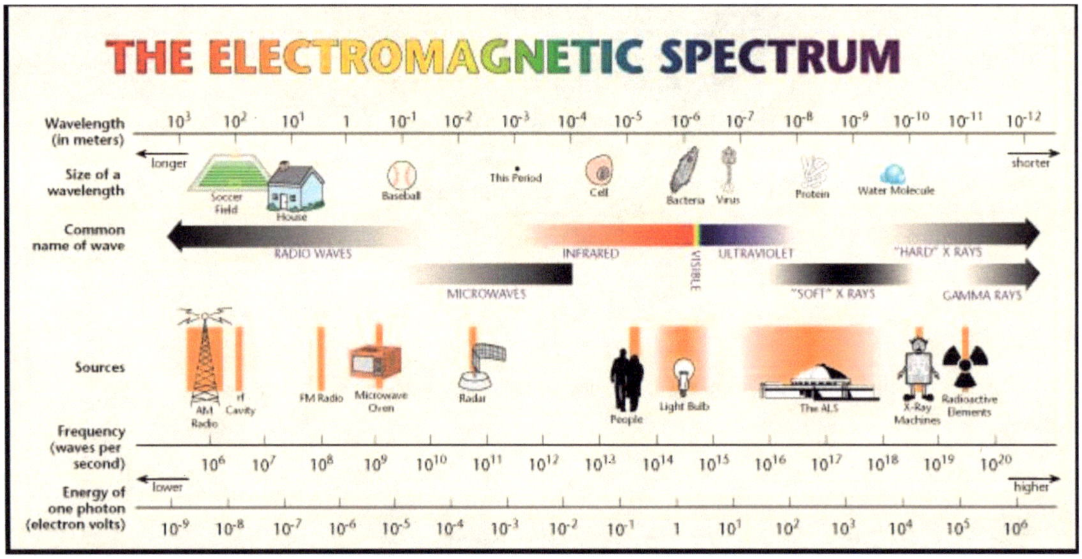

Source: https://en.wikipedia.org/wiki/Gamma_ray#/media/File:Cont_emspec2.jpg

THE RAINBOW

We know that white light is naturally occurring in the cosmic and gamma wave frequency bands:

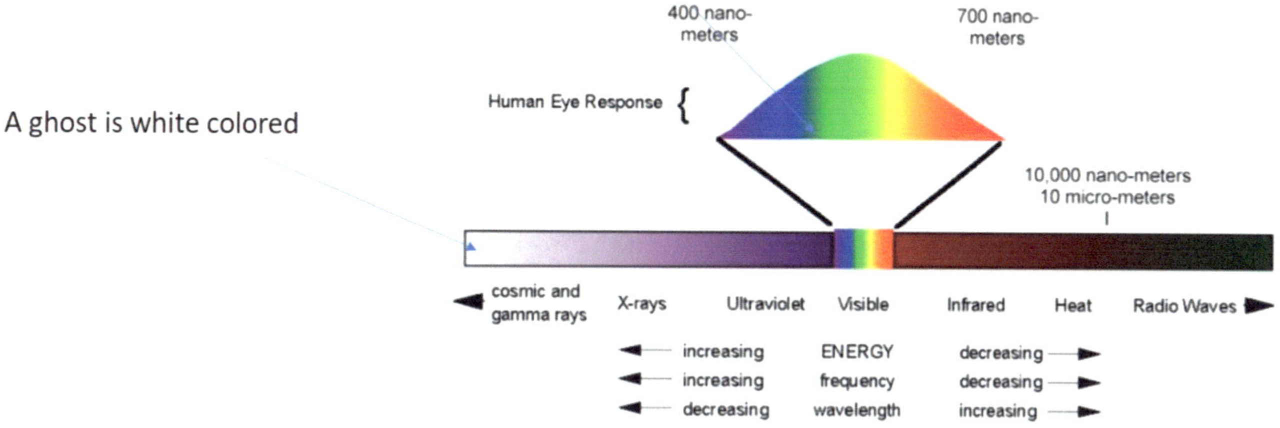

The visible spectrum or color spectrum is a subset of the electromagnetic spectrum. The electromagnetic spectrum is a range of frequencies of different energy waves such as gamma rays, X rays, ultraviolet rays, visible light, infrared waves, microwaves and radio waves. The visible light frequencies lie between the frequencies of the ultraviolet rays and infrared waves. Since the ghost is white almost 100% of the time we can assume that when we manufacture electronic paranormal investigation electronic gadgets that are designed to detect ghosts in the electromagnetic spectrum that they exist in. Our devices would also detect or examine how, how long and under what conditions the ghost pre-manifests and post manifests.

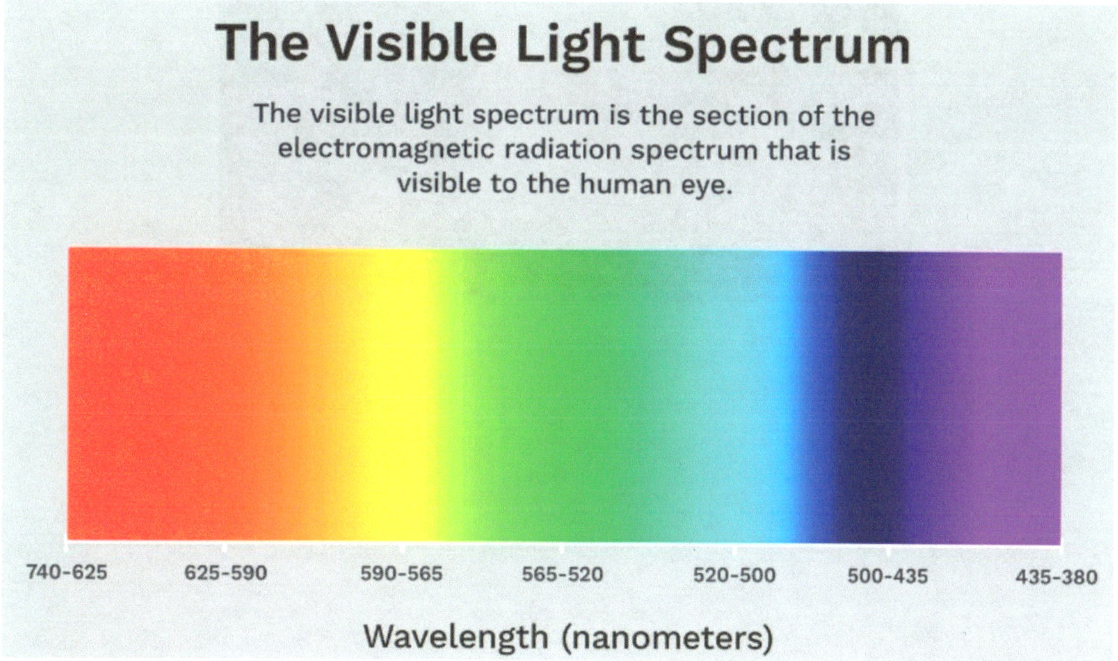

Source: https://www.thoughtco.com/the-visible-light-spectrum-2699036

THE LIGHT BULB

We know that white light is possible with high temperatures of incandescent light. The hotter the object, the whiter it gets (thermal reaction). We know that some people get the chills and the room gets colder. This means that the temperature of a haunted room was lowered suddenly, indicating some kind of thermal reaction.

CLOUDS

We know that clouds are white:

Image credit: NASA/JPL-Caltech/Alex Novati

Heat causes some of the liquid water – from places like oceans, rivers and swimming pools – to change into an invisible gas called water vapor. This process is called evaporation and it's the start of how clouds are formed.

Source: https://climatekids.nasa.gov/cloud-formation/#:~:text=Clouds%20are%20created%20when%20water%20vapor%2C%20an%20invisible,like%20dust%2C%20that%20are%20floating%20in%20the%20air.

Might it not be coincidence that just like when clouds form some paranormal entities appear when humans feel warm, then cold, or a room suddenly feels cold, or humans get chills?

THE MEDICAL X-RAY & TUNGSTEN

The color white also appears in a medical X-ray:

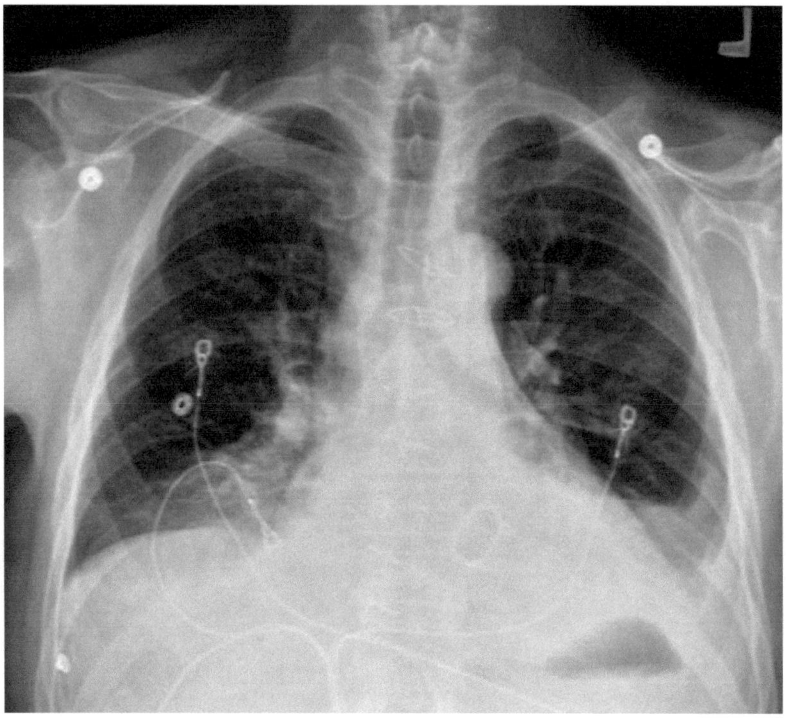

Source: https://radiopaedia.org/images/26385286

An X-ray is typically produced in an X-ray tube by accelerating electrons through a potential difference. It then directs them to a target material, such as tungsten.

Tungsten, just like a ghost, is rare but is found all over the Earth. It's a component in light bulbs, and our paranormal friends love to make our lights flicker! Tungsten is a catalyst. A catalyst speeds up chemical reactions that wouldn't normally happen and makes it happen quicker. Catalysts are not chemically altered during a reaction, so the catalyst can be used repeatedly to create a desired outcome. The use of tungsten in light bulbs is consistent with the behavior of a ghost. Normally, while interacting with the ghost lights flicker, batteries lose their charge, televisions turn on or off and generally electronic things go haywire. I believe that the paranormal entity is extracting the tungsten properties of these items to generate energy to manifest to humans.

Tungsten is a biomolecule that has properties of Morphogenesis. Morphogenesis can take place in a mature organism, such as in the normal maintenance of tissue by stem cells or in regeneration of tissues after damage. Morphogenesis also describes the development of unicellular life forms that do not have an embryonic stage in their life cycle. Morphogenesis is essential for the evolution of new forms. Morphogenesis is a mechanical process involving forces that generate mechanical stress, strain, and movement of cells, and can be induced by genetic programs according to the spatial patterning of cells within tissues. (Source: Wikipedia.org)

CHARGE IT

Any of the alkali metals (the first group in the periodic table) will react pretty violently with water and cause a fire or explosion.

Lithium and other batteries typically lose charge while a paranormal investigation team is performing an investigation. The entity may be using the lithium battery as a source of energy to manifest.

THE MIRROR

This morning you probably looked in the mirror before you left the house for work. Tonight, you may look into the mirror when you brush your teeth. The dance floor has a disco ball hanging and spinning over your head while you dance. Mom has a mirror in her purse. There is a mirror in the restroom and on our cars. There are mirrors in department stores to try on clothes.

Yes. My paranormal clients have told me that the paranormal do use mirrors as portals, and in a few cases, I studied two mirrors facing each other when the paranormal entity was occupying a home. Painting the mirror frame, sanding it, swapping it out, removing it, or replacing it influences the paranormal activity attached to it.

Silver, aluminum, gold!

There are so many cultures that recommend how to dispose of a mirror. The internet is full of mirror destruction methods. I'll let those speak for themselves.

Mirrors are special to the paranormal because it's a method of transportation.

There are chemicals in the mirror that prevent the mirror from rusting (a.k.a. decomposing).

Here is a connection between facts and science right in front of us. No one notices it.

It really confuses me as a paranormal investigator why so many investigation teams don't bring mirrors with them if the location they are investigating doesn't have mirrors.

I am also confused but will mention it to the manufacturers of paranormal investigation equipment.

If we know for a fact that a mirror is a portal we can create commercial paranormal detection devices with the same materials used to make a mirror.

There is something in the materials of the mirror that the paranormal entity can use to travel.

We know what a mirror is, the paranormal knows what a mirror is. Most likely the paranormal entity knows what the mirror is made of, so do humans. We both know the paranormal can travel through it.

Just so we're perfectly clear. A paranormal entity is using a human made mirror to travel between dimensions.

I'm just as guilty as you. I take the mirror for granted.

PEARLS

Natural white pearls are nearly 100% calcium carbonate and conchiolin. Conchiolin is a protein released by the mollusk.

For the rest of this book we'll review manmade or natural substances or processes that generate their own light, make the color white, or may allow a ghost to extract energy from a source to manifest, and examine phenomenon relating to how a ghost manifests.

HOST/GUEST CHEMISTRY

Host–guest chemistry describes complexes that are composed of two or more molecules or ions that are held together in unique structural relationships by forces other than those of full covalent bonds. The deceased person is the host and the ghost is the guest.

PRECIPITATES

A precipitate is a solid material that emerges from a liquid solution by chemical or physical change. It settles to the bottom of the container or can be separated by filtering or suspension amorphous, or crystalline.

AMORPHOUS

A ghost is an amorphous entity in that it lacks a clear structure of focus and has the general characteristics of a human shape.

BATTERY TERMINAL CORROSION

The sulfuric acid within the battery releases hydrogen gas, which causes battery deterioration. As the gasses react with the surrounding atmosphere, a corrosive environment emerges. Moisture and salts, on the other hand, just speed up the process. A car battery uses negative and positive electrical energy, and coincidentally car batteries and other batteries in electronic devices do lose power or the battery drains during a paranormal encounter.

STRONTIUM

WHITE TO BLACK COLOR CHANGER

An earth metal, strontium is a soft silver-white yellowish metallic element that is highly chemically reactive. The metal forms a dark oxide layer when it is exposed to air. Strontium has physical and chemical properties similar to those of its two vertical neighbors in the periodic table, calcium and barium. It occurs naturally mainly in the minerals celestine and strontianite, and is mostly mined from these.

https://en.wikipedia.org/wiki/Strontium

SILVER PHOSPHATE OR SILVER ORTHOPHOSPHATE

A chemical used in photographic production. Silver colored.

BARIUM SULFATE

Just like a ghost this has a white color. It is used for x-ray imaging and x-rays are white. The “barium meal”. It’s used in white paints. It is and was used in photographic papers to diffuse light. It is highly reflective.

TAPHONOMY

Taphonomy, from the Greek, taphos, meaning tomb or grave, and nomy, meaning classification, is a field of paleontology, paleo-anthropology, and bioarchaeology that studies human and animal remains in relation to the post-mortem (after death) transformations that occur in burial sites.

https://www.encyclopedia.com/science-and-technology/biology-and-genetics/biology-general/taphonomy.

STROBE LIGHTS (CFF)/WAGON WHEEL EFFECT

The flicker fusion threshold, critical flicker frequency (CFF) or flicker fusion rate, is a concept in the psychophysics of vision. It is defined as the frequency at which an intermittent light stimulus appears to be completely steady to the average human observer.

For the purposes of presenting moving images, the human flicker fusion threshold is usually taken between 60 and 90 Hz or higher.

Movies are recorded at 24 frames per second and displayed by repeating each frame two or three times for a flicker of 48 or 72 Hz.

Standard-definition television operates at 25 or 30 frames per second, or sometimes at 50 or 60 (half-) frames per second through interlacing.

High-definition video is displayed at 24, 25, 30, 60 frames per second or higher.

https://en.wikipedia.org/wiki/Flicker_fusion_threshold

COLORIMETRY

The study of how the human eye receptors and how they interpret colors.

APLASTIC ANEMIA

When a celled organism can't fully manifest due to development problems. Aplastic anemia may be the default structure of a host.

ANAPLASIA

A condition of cells that allow them to multiply.

NEOPLASIA

The abnormal growth of cells.

HYPERPLASIA

The incomplete development of tissue or organisms. The human body gets "puffy" after passing away. The cells grow, but then eventually die.

METAPLASIA

The transformation of one differentiated cell type to another differentiated cell type. The change from one type of cell to another may be part of a normal maturation process, or caused by some sort of abnormal stimulus. In simplistic terms, it is as if the original cells are not robust enough to withstand their environment, so they transform into another cell type better suited to their environment. If the stimulus causing metaplasia is removed or ceases, tissues return to their normal pattern of differentiation. When cells are faced with physiological or pathological stresses, they respond by adapting in any of several ways, one of which is metaplasia. (Wikipedia.org)

EPIGENETICS

The study of stable changes in cell function (known as *marks*) that do not involve alterations in the DNA sequence. This is part of how humans inherit traits from our parents or grandparents or ancestral lineage. (Hair, eyes, facial features, height, cancer, heart disease) and other human aspects that occur as the family tree grows. In epigenetics only the human shape and form and exterior physical features of the human and apparel appear to the human. All other features are silenced. This mechanism enables differentiated cells in a multicellular organism to express only the genes that are necessary for their own activity. So if a ghost appears of its own free will, or by some other external catalyst then the ghost has a specific purpose and the minimum amount of cells needed to achieve its purpose is by being white, machine the DNA of a human and appearing to the living. Where epigenetics involves DNA it would corroborate that humans smell sweet scents when ghosts appear, which is consistent with carbon tetrachloride and trichloroethylene.

PLAST

A small body, structure, particle, or granule, especially of living matter.

APLASIA DEFECT

An organ or tissue is wholly or largely absent.

DYSPLASIA

Any of various types of abnormal growth or development of cells (microscopic scale) or organs (macroscopic scale),

TROPHIC MUTUALISM

How energy and nutrients are transferred between species.

ATYPIA

Cells are irregular or abnormal.

ZINC HYDROXIDE

It is used to make surgical dressings that are also white. Many, many paranormal investigations are at abandoned hospitals.

CERUSSITE

The mineral is usually colorless or white, sometimes grey or greenish in tint and varies from transparent to translucent with an adamantine luster (diamond-like shine).

VENETIAN CERUSE

A former lead-based cosmetic cream used to make female skin white. Most commonly associated with Queen Elizabeth 1. The side effect of using this cosmetic is that it removed the outer layers of the skin. A ghost appears as a human or the outer skin of a human with no internal organs.

ZINC CARBONATE

A white compound that adopts the same structure as calcium carbonate (calcite). Calcite is a common substance found in white-colored chalk and limestone, eggshells, gastropod shells, shellfish skeletons, and pearls. Calcium carbonate is the active ingredient in agricultural lime and is created when calcium ions in hard water react with carbonate ions to create lime scale. Lime scale is the “white” film on showers and water pipes. It has medical use as a calcium supplement or as an antacid. Many antacids are white colored.

CALCIUM CARBONATE

A white substance associated with chalk, which is white, marble which tombstones and headstones are made of, and travertine, the white substance found on the edges of hot springs. Used to make white diapers, white paper, white paint, adhesives, toothpaste.

ARAGONITE

The formation of aragonite is promoted by the presence of magnesium ions. Magnesium ions are also present in the firefly or lightning bug.

CHEMICAL KINETICS

Chemical kinetics, also known as reaction kinetics, is the branch of physical chemistry that is concerned with understanding the rates of chemical reactions. It is different from chemical thermodynamics, which deals with the direction in which a reaction occurs but in itself tells nothing about its rate. Chemical kinetics includes investigations of how experimental conditions influence the speed of a chemical reaction and yield information about the reaction's mechanism and transition states, as well as the construction of mathematical models that also can describe the characteristics of a chemical reaction. Source: https://en.wikipedia.org/wiki/Chemical_kinetics

ELECTRICITY AND SWEET SMELLS

Iron conducts electricity and iron is in human blood. We know that DC current electricity flows from negative to positive. We also know that paranormal entities make lights flicker before, during or after an interaction with the living. We also know that car batteries and cell phone batteries drain quickly during paranormal encounters.

It is proven science that human blood has an electrical charge, which may indicate that the paranormal entity draws energy from the human's blood to get energy to interact with the living. This may be why humans get the chills, or their body hair stands up or they feel cold spots on their body, or even why the room goes cold. This indicates a thermal reaction or a biological reaction.

The human body contains an anticoagulant called heparin. Heparin does not allow the blood to coagulate. Heparin has a very high negative charge density. Heparin is produced in the body by the basophils and mast cells. Blood thus becomes negatively charged due to presence of negatively charged heparin in it. Thus, blood is a negatively charged sol. The fluid used for dispersion possesses a charge opposite to that of the colloidal particles. Thus, the charges are balanced and the colloidal solution becomes electrically neutral. The stability of a colloidal solution depends on the charge possessed by it. Colloidal solutions have very minute particles and they are not visible by naked eyes. If the fluid used for dispersion of colloids is water then the sol is known as hydrosol. If the fluid used for dispersion of colloids is air then the sol is known as aerosol. Hemoglobin in blood is positively charged. It is inside the red blood cell. The membrane of the red blood cell is negatively charged. Examples of positively charged sols: Hemoglobin, metal hydroxides, basic dyes, etc. Examples of negatively charged sols: Metal sols, metal sulfides, gold sols, blood, starch, acidic dyes, etc. We also know that human blood is about 1 degree warmer than the temperature of the human body and this is done both through the liquid part of the blood (plasma), which can absorb or give off heat, as well as through the speed at which the blood is flowing: When the blood vessels expand, the blood flows more slowly and this causes heat to be lost. When the temperature outside the body is low, the blood vessels can contract to reduce the amount of heat lost. We also know that some humans smell sweet or citrus scents when they encounter paranormal entities. Hydrosol is used for perfume, and aromatherapy.

https://www.vedantu.com/question-answer/blood-is-a-charged-sol-a-positively-b-neutral-c-class-12-chemistry-cbse-5f8cf79946566e310be95a66

https://www.healthline.com/health/hydrosol#overview

https://www.ncbi.nlm.nih.gov/books/NBK279392/

POLYMORPH

A polymorph is an organism or inorganic object or material which takes various forms.

SPATIAL ISOMERISM/STEREOCHEMISTRY/STEREOCHEMISTRY

In stereochemistry, stereoisomerism, or spatial isomerism - molecules have the same molecular formula and sequence of bonded atoms (constitution), but differ in the three-dimensional orientations of their atoms in space. This contrasts with structural isomers, which share the same molecular formula, but the bond connections or their order differs. By definition, molecules that are stereoisomers of each other represent the same structural isomer.

IRREVERSIBLE THERMODYNAMICS

Irreversible thermodynamics is a division of physics which studies the general regularities in transport phenomena (heat transfer, mass transfer, etc.) and their relaxation. I believe that since a ghost appears and disappears that the reason it disappears is because it doesn't have enough energy to sustain itself, otherwise it would.

METASTABILITY

We can assume that since the ghost appears as a human being that its DNA was human. Some see a male ghost who looked like the male owner of a house hundreds of years ago. Others see female's ghosts who wear dresses.

A ghost has a finite lifetime when it appears to humans.

In isolation metastability is the state of least energy is the only one the system will inhabit for an indefinite length of time, until more external energy is added to the system (unique "absolutely stable" state); - Consciousness of the ghost

The system will spontaneously leave any other state (of higher energy) to eventually return (after a sequence of transitions) to the least energetic state- the ghost extracts energy to manifest, then disappears.

The clothing of a ghost must be relevant to its existence or the ghost would not manifest the clothing, the ghost doesn't know its manifesting clothing, or if it's manifesting clothing it might not need to or want to manifest the human body that would be underneath the clothing.

https://en.wikipedia.org/wiki/Metastability#Statistical_physics_and_thermodynamics

RUBIDIUM CHLORIDE

Rubidium chloride is involved with negatively charged expanding DNA. It's a non-invasive biomarker. A biomarker or biological marker, is a measurable indicator of some biological state or condition, especially after some kind of medical treatment. It's a baseline. A ghost needs a baseline of some sort (DNA) to manifest in the shape and appearance of a human. Some biomarkers have powerful implications if used for testing, so much that sometimes the patient must sign waivers or permissions to undergo testing.

The term biomarker refers to changes in processes of living organism, some of which includes the results that occur when an organism is irritated or exposed to substances or environments that cause irritation.

This alkali metal halide salt is composed of rubidium and chlorine, and finds diverse uses ranging from electrochemistry to molecular biology. We know that paranormal entities cause electricity and batteries to act strangely and we know that a ghost is the same shape and form of a biological human.

In theory a ghost could simply be experiencing a chemical or biological irritation that is quickly resolves, then the ghost disappears.

METEORS

I'm not suggesting that meteors are the cause of paranormal entities and ghosts or that ghosts are aliens from other planets. We do know, scientifically, that the contents and make up of the meteor that is burning up as it heads to the earth's surface determines the color of the meteor. We can assume that the chemical composition of the meteor, if that chemical is also found on earth, that a ghost may use those chemicals and be the color of that burning chemical.

FIREWORKS

I'm not suggesting that ghosts appear as a result of a firework on the 4th of July, but these chemicals, singly or combined when ignited cause certain colors to be visible to the naked human eye:

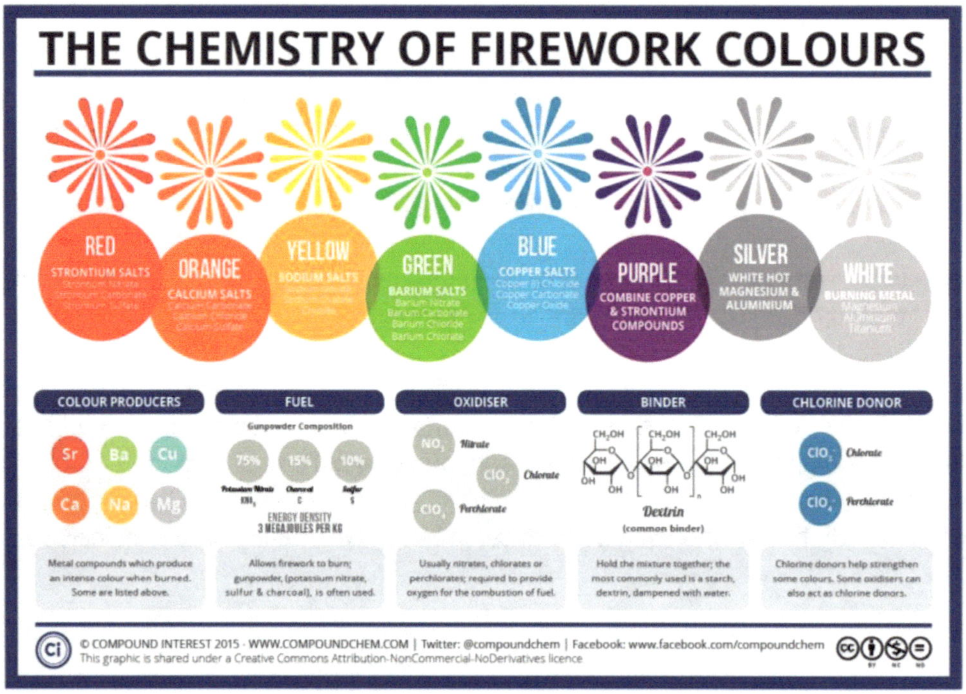

https://earthsky.org/human-world/how-do-fireworks-get-their-vibrant-colors/

BIOLUMINESCENCE

Bioluminescence is the ability of an organism to produce its own light. We can assume that a ghost does this as well.

Just like a ghost, a lightning bug emits a low level of light that quickly turns on and off. My theory of ghosts not containing ATP comes to a screeching halt when compared to how a lightning bug creates light.

In the body of the lightning bug the enzyme luciferase acts on the luciferin in the presence of magnesium ions, a chemical called adenosine triphosphate (ATP) and oxygen to produce light. The light that some fireflies produce is extremely efficient. In fact, it's the most efficient light in the world! Nearly 100% of the chemical reaction's energy becomes light. The light that fireflies produce may be green, yellow or orange in color.

The cells contain a chemical called luciferin and make an enzyme called luciferase. To make light, the luciferin combines with oxygen to form an inactive molecule called oxyluciferin. The luciferase speeds up the reaction, which occurs in two steps:

The luciferin combines with adenosine triphosphate (ATP), which is found in all cells, to form luciferyl adenylate and pyrophosphate (PPi) on the surface of the luciferase enzyme. The luciferyl adenylate remains bound to the enzyme: luciferin + ATP -------------> luciferyl adenylate + PPi

The luciferyl adenylate combines with oxygen to form oxyluciferin and adenosine monophosphate (AMP). Light is given off and the oxyluciferin and AMP are released from the enzyme's surface: luciferyl adenylate + O2 -------------> oxyluciferin +AMP + light

But on the tail end of all this, many of my paranormal clients who had paranormal encounters were taking magnesium supplements. There is a connection between magnesium and the ability of organisms to produce light. Magnesium is also used in modern appliances and flash photography. When magnesium burns it burns the color……white!

Luciferin can be regenerated continuously and light can be produced again. The regeneration occurs in three sequential steps: the oxyluciferin produced from the last bioluminescent reaction generates 2-cyano-6-hydroxybenzothiazole (CHBT) in the luciferin regenerating enzyme (LRE) via a hydrolysis reaction; CHBT combines with L-cysteine in vivo to form L-luciferin via a condensation reaction; and L-luciferin inverts into D-luciferin in luciferase and thioesterase. The presently proposed mechanism not only supports the sporadic evidence from previous experiments but also clearly describes the complete process of luciferin regeneration.

Cheng YY, Liu YJ. Luciferin Regeneration in Firefly Bioluminescence via Proton-Transfer-Facilitated Hydrolysis, Condensation and Chiral Inversion. Chemphyschem. 2019 Jul 2;20(13):1719-1727. doi: 10.1002/cphc.201900306. Epub 2019 Jun 12. PMID: 31090243.

HUMAN BIOLUMINESCENCE

In Japan at the Tohoku Institute of Technology Daisuke Kikuchi and Masaki Kobayashi conducted a study where inside the human body a photo protein combined with an ion of calcium produces light.

The light wasn't visible by the human eye, was brightest in late afternoon and at its lowest late at night. The light was brighter in the forehead, cheeks and neck compared to other body parts.

Source: "Bioluminescence and Bio photons in Humans, by Images Scientific Instruments Inc. 1-800-230-4535 copyright 2007-2023.

Some of my paranormal clients were taking calcium supplements. Another client, who admitted she regularly hosted overnight ghosts as they traveled through her town, reported that the ghosts liked it when she drank orange juice because the ghost received a burst of energy when she drank it.

Orange juice contains calcium.

Also, some paranormal clients report smelling sweet scents for both ghosts and shadow figures. However, for shadow figures the smell was more like dirt.

GLOW STICK

Diethyl Prithalate (DP), Phenylethynl anthrarene (green dye) and ribrene (yellow dye) mixed together to make a white colored glow stick. Mixed with 2, 4, 6 trichlorophenyl (the stuff that glows) and sodium acetate or baking soda and a dash of hydrogen peroxide.

Glow stick built.

A chemical reaction to create white light that is temporary. Is the ghost doing the same thing in a smaller quantity?

EXHALE

Like a ghost that only appears for a few seconds, so does the mist coming out of our mouths in the cold weather when we exhale.

Our exhaled air contains Argon. Argon is the most abundant gas on Earth. It's used in burning and light. It doesn't form any stable compounds at room temperature. Argon forms clathrates, which are chemical substances that trap or collect molecules.

FRACTIONAL DISTILLATION

The separation of a mixture into its component parts, or fractions. Chemical compounds are separated by heating them to a temperature at which one or more fractions of the mixture will vaporize. (Wikipedia).

THE HIGHWAY

It's a summer day and you're driving down the Interstate. Looking ahead there are "puddles" of water on the road that appear and sometimes disappear. When you get to the stop on the road where you saw the puddle the puddle isn't there.

HYDROGEN PEROXIDE

When hydrogen peroxide is placed on the cut of the human skin the peroxide bubbles a white color then disappears. A ghost appears for the same duration, then disappears.

PHOTO PROTEIN

In biochemistry, any of several proteins that give off light upon combination with oxygen, hydrogen peroxide, or other oxidizing agents. Unlike the oxidation of luciferin, the production of light by a photo protein requires no catalyst. Such a system occurs in Aquaria, a luminescent jellyfish, in which the single organic chemical aequorin requires calcium or strontium ions for luminescence to occur. Humans ingest calcium supplements, calcium is found in milk and bones.

https://www.britannica.com/science/photoprotein

Signs of a catalyst where a ghost may have to extract energy from the Earth or other sources external to the ghost to manifest include lights flickering, electronic gadgets, human chills, and room temperatures dropping.

If these paranormal encounters are not catalysts, then we can assume that a ghost is a photo protein.

QUANTUM OPTICS

The study of how individual's quanta of light (photons) interact with protons & molecules.

It was first assumed by Plato that human beings saw color because their eyes alone enabled them to see color.

BLACK BODY

I mentioned a shadow person in a previous chapter. A black body or blackbody is an idealized physical body that absorbs all incident electromagnetic radiation, regardless of frequency or angle of incidence. The name "black body" is given because it absorbs all colors of light.

Since a shadow person is all black, including its hat and bow tie it may be a black body. However, some shadow people have colored eyes.

Human beings can actually create shadow people, especially if their music and audio headphones go haywire.

According to Neurologist Olaf Blanke of the Brain Mind Institute in Lausanne, Switzerland, and his colleagues, they were attempting to identify the source of epileptic seizures in a 23-year-old woman. They applied a mild current through surgically implanted electrodes to various regions of her brain. Not much happened until the researchers stimulated the woman's left Temporoparietal Junction (TPJ), located roughly above the left ear.

Suddenly, she reported feeling the presence of a mystery person behind her, a motionless and speechless shadow that imitated her body posture and actions. "He" lay beneath her when she lay down, sat behind her when she sat down, and attempted to take a test card from her when she tried to participate in a language exercise.

Source: https://www.science.org/content/article/illuminating-shadow-people?cookieSet=1

Source: https://everipedia.org/wiki/lang_en/Temporoparietal_junction

angular gyrus is the spot on the brain Source: Pinterest.com

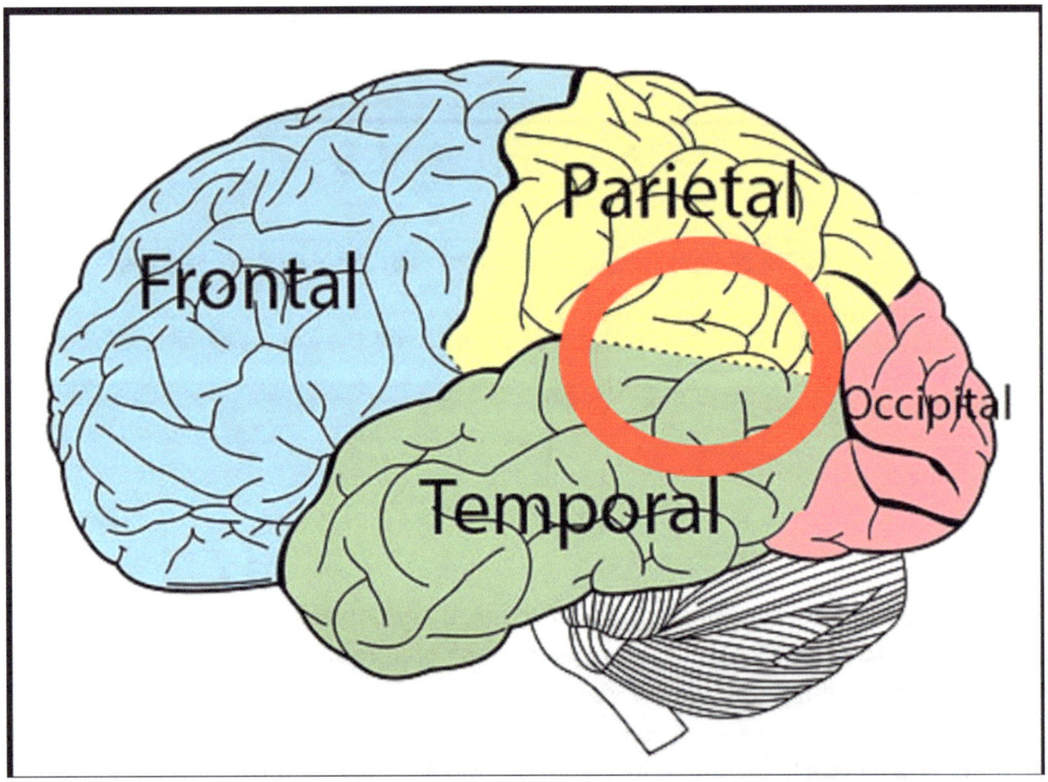

THE FIRST LAW OF THERMODYNAMICS:

Energy can be transformed from one form to another, but can be neither created nor destroyed.

THE FOG MACHINE

This cool device uses Propylene glycol and Triethylene glycol.

Just like a ghost, these two chemicals have a sweet scent.

Propylene glycol is associated with lactic acid, coffee products, soda, skin lotion, pharmacy products, and e-cigs. All of which would be inside the human body at time of death. It is also used with DNA preservation. It prevents water from becoming puddles (maintains water mist)

Triethylene glycol - Absorbs water from the environment, nail polish, vinegar based, flammable, dehydrates natural gas, is involved with infections of the human skin, it doesn't need oxygen to function and is found in hospital infections, especially with catheters. It's o amazing that the paranormal community investigates hospitals all the time!

THE PERIODIC TABLE OF ELEMENTS

If we assume that ghosts are white and we want to incorporate science into the ghost then one of the most visible signs of human science is the periodic table of elements. Upon examination, there are no elements that are the color white. There are elements that are colored silver, gray and slate gray. It may be that a ghost is actually silver, gray or slate gray but the ghost appears as white colored when it's set in a dark background, such as night.

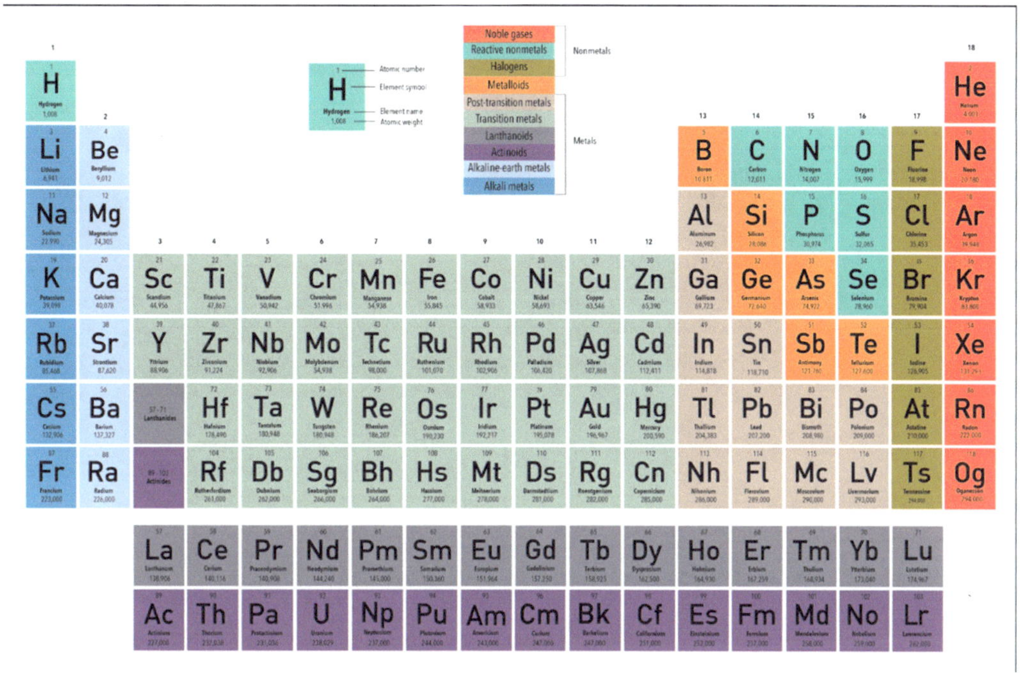

Gray Colored Elements

Atomic Number	Symbol	Name
14	Si	Silicon
26	Fe	Iron
27	Co	Cobalt
28	Ni	Nickel
32	Ge	Germanium
34	Se	Selenium
41	Nb	Niobium
42	Mo	Molybdenum
72	Hf	Hafnium
73	Ta	Tantalum
74	W	Tungsten
75	Re	Rhenium
78	Pt	Platinum
83	Bi	Bismuth

Silver Colored Elements					
Atomic Number	Symbol	Name	Atomic Number	Symbol	Name
31	Ga	Gallium	59	Pr	Praseodymium
3	Li	Lithium	60	Nd	Neodymium
11	Na	Sodium	61	Pm	Promethium
12	Mg	Magnesium	62	Sm	Samarium
13	Al	Aluminium	63	Eu	Europium
19	K	Potassium	64	Gd	Gadolinium
20	Ca	Calcium	65	Tb	Terbium
21	Sc	Scandium	66	Dy	Dysprosium
22	Ti	Titanium	67	Ho	Holmium
23	V	Vanadium	68	Er	Erbium
24	Cr	Chromium	69	Tm	Thulium
25	Mn	Manganese	70	Yb	Ytterbium
33	As	Arsenic	71	Lu	Lutetium
37	Rb	Rubidium	77	Ir	Iridium
38	Sr	Strontium	80	Hg	Mercury
39	Y	Yttrium	81	Tl	Thallium
40	Zr	Zirconium	84	Po	Polonium
43	Tc	Technetium	85	At	Astatine
44	Ru	Ruthenium	87	Fr	Francium
45	Rh	Rhodium	88	Ra	Radium
46	Pd	Palladium	89	Ac	Actinium
47	Ag	Silver	90	Th	Thorium
48	Cd	Cadmium	91	Pa	Protactinium
49	In	Indium	92	U	Uranium
50	Sn	Tin	93	Np	Neptunium
51	Sb	Antimony	94	Pu	Plutonium
52	Te	Tellurium	95	Am	Americium
55	Cs	Cesium	96	Cm	Curium
56	Ba	Barium			
57	La	Lanthanum			
58	Ce	Cerium			

Slate Gray Elements		
Atomic Number	Symbol	Name
4	Be	Beryllium
30	Zn	Zinc
53	I	Iodine
76	Os	Osmium
82	Pb	Lead

FAREWELL

There are many other chemicals and naturally recurring phenomenon that glow or turn white or give off their own light. I believe that since a ghost occupies the Earth that it is subject to the laws of physics on Earth and therefore we can predict its behavior, and prevent it from manifesting to harm humans. We already know they exist. It's time to move forward. As a profession, paranormal investigators and ghost hunters can't keep disagreeing if an orb is in a photo or video or not. We can't keep posting videos and photos on social media just for fun. Our profession will be regulated by legislators soon. We have to focus on tools, techniques, processes, inputs and outputs that help us prevent harmful entities from harming the living. Anything else is a waste.

Thanks for reading. Some of the greatest inventions, achievements and monumental discoveries and events ever made in human history was from the person who was an over thinker.

If you have feedback on this book (positive or negative), I welcome it at: kirk_mcmanus@yahoo.com.
Godspeed.

Made in the USA
Monee, IL
05 November 2026

67807093R00066